FRANK KINGDON-WARD

By the same author

Sir Joseph Banks
The Plant Hunters
The Kitchen Garden
The Herb Garden

FRANK KINGDON-WARD

The Last of the Great Plant Hunters

Charles Lyte

JOHN MURRAY

© Charles Lyte 1989
First published 1989
by John Murray (Publishers) Ltd
50 Albemarle Street, London W1X 4BD

All rights reserved
Unauthorised duplication contravenes applicable laws

Typeset and printed by
Butler & Tanner Ltd, Frome and London

British Library Cataloguing in Publication Data

Lyte, Charles, 1935–
Frank Kingdon-Ward: the last of the great plant hunters.
1. Plants. Collecting. Kingdon-Ward, Frank
I. Title
579'.6

ISBN 0–7195–4735–0

This book is dedicated to Jean Rasmussen, who after her marriage to Frank Kingdon-Ward accompanied him on all his post-war expeditions. Without her enthusiasm, help, generosity, patience and friendship this biography would not have been possible.

Contents

Illustrations	viii
Acknowledgements	ix
Introduction	xiii
Chapter 1: Childhood and Youth	1
Chapter 2: First Steps into the Wilderness	13
Chapter 3: The Making of a Plant Hunter	29
Chapter 4: Back to the Flower Trail	59
Chapter 5: To the Land of the Great Gorge	69
Chapter 6: Kingdon-Ward Territory	95
Chapter 7: From the Thames to the High Himalaya	115
Chapter 8: Plain Mr Ward	139
Chapter 9: Of Tea, and 'Planes, and Lilies	163
Chapter 10: Earthquake	181
Chapter 11: The Last Lily	193
Expeditions	207
Books by F. Kingdon-Ward	209
Bibliography	211
Index	213

Illustrations

Between pages 112 *and* 113
1. Professor Harry Marshall Ward
2. Frank sitting on his mother's knee
3. Frank and Winifred
4. Frank with Kenneth Ward on their bicycles
5. Winifred in her late teens
6. Frank in Shanghai
7. The young explorer
8. In Mesopotamia, 1917
9. *Rhododendron nuttallii*, Adung Valley
10. Setting off on the plant trail
11. K-W crossing the Ata Chu
12. Tibetan minstrels
13. *Primula agleniana*
14. *Lilium nepalense*
15. Orange-flowered dendrobrium
16. *Lilium wallachianum*
17. *Stauropsis polyantha*
18. *Michelia doltsopa* in the Adung Valley
19. Collecting in North Burma
20. K-W and dendrobrium orchid
21. With Jean outside their camp
22. *Lilium mackliniae*
23. Jean pressing plant specimens
24. Base camp in North Burma
25. Mishmi porters carry Jean to safety
26. Frank Kingdon-Ward

All the botanical photographs were taken by K-W himself in the wild. Photographs nos. 9, 11, 12, and 18 are reproduced by courtesy of the Royal Geographical Society: no. 26 is reproduced by courtesy of Popperfoto.

Acknowledgements

THE existence of this biography is in great measure due to the unstinting help and generosity of many people, to whom I am extremely grateful: Mrs Jean Rasmussen for access to diaries and letters and for permission to quote from all Kingdon-Ward's writings; Mrs Pleione Tooley (*née* Kingdon-Ward) and Martha Kingdon-Ward for their time, and for lending letters and documents; Sir George Taylor for advice and letters; Lord Cawdor for permission to quote from his father's Tsangpo diaries; Mrs Elizabeth Barraclough; Mrs Anne Smith; Mrs Lochlan; Mrs Ruth Annesley; Lisa Radcliffe, for heroic typing; Mr Rudolf Ziesenhenne; Lord Cranbrook; Mr Ronald Kaulback; Mrs R. L. Yeates; Mike Park and Ian Smith, gardening and natural history book specialists; Mrs P. C. Grigg of the New Zealand Rhododendron Association; Mr and Mrs Eric Chester; the staff of the Royal Botanic Gardens Library, Kew, especially Sylvia FitzGerald, Chief Librarian and Archivist, Leonore Thompson, Archivist, and Marilyn Ward, Illustrations Assistant; Dr Brent Elliott, Chief Librarian of the Royal Horticultural Society Lindley Library, and his staff; Tony Schilling, Deputy Curator, Wakehurst Place; Dr E. Charles Nelson, Botanist at the National Botanic Garden, Dublin; and the High Master and staff of St Paul's School. My particular thanks to Dr John Hemming, Director and Secretary of the Royal Geographical Society; David Wileman, RGS Chief Librarian, and his assistant librarians Edwin Trout and Jayne Dunlop; and RGS Picture Librarian Nick Howarth.

EXPLORATION IN THE EASTERN HIMALAYAS AND THE RIVER GORGE COUNTRY OF SOUTHEASTERN TIBET
Frank Kingdon Ward (1885–1958)

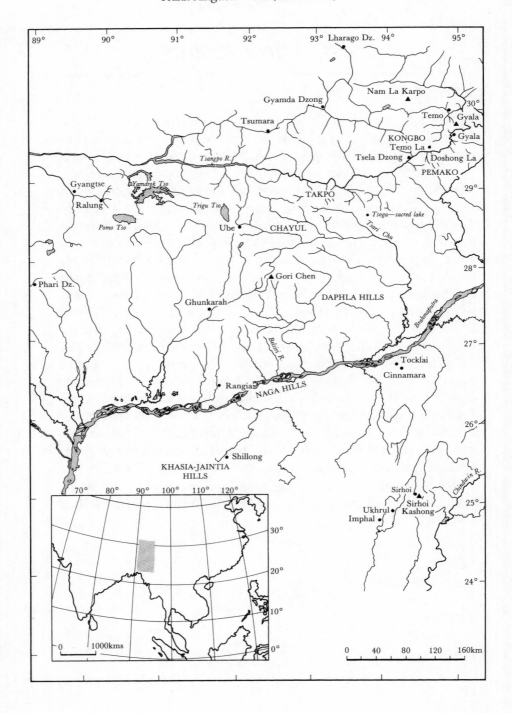

EXPLORATION IN THE EASTERN HIMALAYAS AND THE RIVER GORGE COUNTRY OF SOUTHEASTERN TIBET
Frank Kingdon Ward (1885–1958)

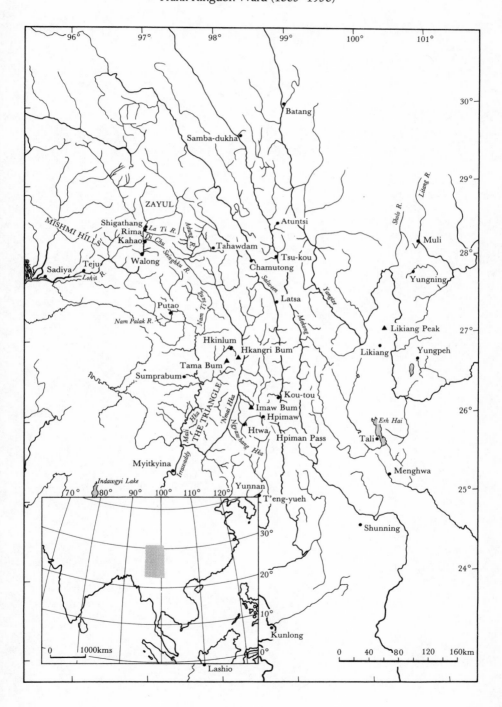

Introduction

FRANK Kingdon-Ward's death in 1958 brought to an end an extraordinary era that had spanned three hundred years; an era of brave and dedicated men who had devoted their lives to the search for plants, from forest giants to alpine dwarfs; plants that brought a new dimension of beauty into human lives; plants that enriched the economies of nations; plants that produced drugs to bring hope and health to the sick. What these men had in common was a singleness of purpose that seemed to inure them from considerations of danger and discomfort.

Certainly no plant hunter ever became a rich man, except in spirit and achievement. So what on earth was it that induced them to surrender all the normal comforts of family life, a steady job, good food, and sound shelter, for danger, disease, loneliness, punishing extremes of climate, physical threats from suspicious local peoples, and a livelihood that ranged from small to nil?

In K-W's case the answer almost certainly lies in a magnificent lifelong, passionate obsession with the wilderness. The vast ranges of the Himalaya, the valleys scooped out by ancient glaciers, the deep gorges, towering cliffs, tumbling screes, the rivers — sinuous and green one moment, the next a roaring, foaming mass of energy — great forests and jungles, and alpine meadows glittering with flowers, were a fascinating Pandora's Box waiting to be unlocked to release not a catalogue of human ills, but a cascade of natural beauty, and perhaps answers to some of the riddles of the making of our planet.

Of course it would be quite wrong to give the impression that he was a starry-eyed romantic drifting dreamily from one flowery mead to another. He was a fine scientist with a sharp and penetrating analytical mind. After taking part in the Bedford Expedition in 1910–11, he formed an ambition to become a pure explorer, discovering and mapping new territories, and to develop and prove theories on the formation of the Earth's crust. But pure exploration was less of a

Introduction

paying proposition than plant hunting. Botany he perceived as a secondary occupation, one that would pay the exploring bills, which was why he was so ready to accept the offer to become a collector for Bees' Nursery.

While he was justifiably awarded the highest honour of the Royal Geographical Society, the Founders' Gold Medal, and the Livingstone Gold Medal from the Royal Scottish Geographical Society, he did not reach the highest ranks of geographer-explorers. Surveying and mapmaking were tasks that he found irksome, and it was clearly a relief to him on a few of his expeditions between the wars to hand the task over to travelling companions. After the Second World War he dropped the geographical side of his work altogether.

K-W's true genius lay in botany and the discovery and collection of plants. In this he was outstanding. He made valuable collections of economic plants – not one of the most glamorous aspects of botanical exploration – and did useful work in searching for species of wild tea to be used in hybridising programmes for commercial strains. But it was in the field of horticultural collecting that he really triumphed. Although he did not paint, he had an artist's eye for floral beauty and, more important, an unerring instinct for an outstanding garden plant. He could sweep a mountainside, or, from some craggy perch, the canopy of a rhododendron forest, with his travel-worn binoculars, and spot something really good beckoning him from among the foliage or the rocks.

A great part of his collecting success was due to meticulous and painstaking searching. Nothing seemed to escape his eye. He did not care if this slow scanning of the undergrowth enraged some of his travelling companions, like Lord Cawdor. His patience in the hunt was heroic. However tired he was after a long day's march, if he spotted something at a distance that excited him, he would immediately set out to locate it before darkness fell. The discovery of the Tea Rose Primula is a typical example. And if for some reason he could not collect a plant, such as at his first sighting of the epiphytic lily, *Lilium arboricola* – it was high in an unclimbable tree – he argued that there must be more specimens around and he would eventually track them down. K-W was also blessed with an extraordinary memory, so precise that he could re-locate a plant at seed-gathering time, even when it was buried by snow.

Introduction

Unlike other great plant collectors such as Joseph Hooker and George Forrest, he did not train up teams of native collectors to trawl for huge quantities of seeds, bulbs, living plants and herbarium specimens. He supervised every stage of collecting himself, from discovery to harvest. There were clearly two reasons for this. The first was the sheer excitement of finding something new and outstanding; the other, more practical, was that he liked to see the plant growing, so that he could make a more accurate judgement of its garden quality than might otherwise have been possible. It also enabled him to make accurate field notes that proved invaluable to the plantsmen and horticulturists who had the job of raising his finds for garden use.

What undoubtedly marked him as a great garden plant collector was his absolute confidence in the quality of the plants he discovered. He refused to be deterred when the first collected specimens of *Lilium mackliniae* to be flowered in gardens were written off as an uninteresting 'dirty white'. He knew that was due to an aberration, and he was proved right.

Gardens throughout the world have benefited from his flair and impeccable taste. From more than twenty expeditions he introduced such glorious plants as the Tibetan blue poppy (*Meconopsis betonicifolia*), the beautifully scented and easily grown pink *Lilium wardii*, and *Primulae florindae* and *bulleyana*, which have settled in Britain and elsewhere as though they were natives. His rhododendron introductions have made an enormous contribution to major collections, and have also brought new blood for hybridising. Two alone would have ensured him a leading place in gardening history, the incomparable yellow-flowered *Rhododendron wardii*, and *R. macabeanum*. Gentians, cotoneaster, berberis and prunus all came in for his careful attention, and we are much the richer for it.

Not all of his plants have remained in cultivation, and some which have are now very rare, but many of them remain vividly alive in his published accounts of his adventures, with their marvellous descriptions of his discovery of those species in the wilderness.

I have known of K-W since my childhood. He was a pre-war friend of my parents, and I heard much about him to stimulate my interest. In this account of his life I hope I have succeeded in going beyond the public image to the man himself.

Introduction

He had extraordinary courage and determination, and a profound sense of duty. Many found him austere, even chill, others knew him as a party-goer, full of fun, even a flirt. He was certainly neither a prig nor a prude. Like many other men he wilted under disappointment, knew bitterness, intense loneliness, even bouts of almost suicidal despair. He overcame all of this, but unlike most of us he had to cope with his private agonies far from home, family, friends and loved ones. He rarely had anybody to turn to for comfort and advice, until, during the last ten years of his life, he found the love and companionship he had always sought with his second wife, Jean. He was in many ways an ordinary man who lived an extraordinary life.

Kingdon was his mother's family name, which he was given as a Christian name. Later in life he decided to incorporate it into his surname by hyphenating it to Ward. Since this was his choice, Kingdon-Ward has been adopted throughout this book.

In his writings, K-W (as I shall refer to him) used capitals for both the generic and specific names of some plants. For the sake of clarity I have used the now more common style of a capital for the generic name and lower case for the specific name, and in the cases where plant names have since changed, I have used his original nomenclature.

1

Childhood and Youth

FRANCIS Kingdon-Ward was born on 6 November 1885 in a house in Manchester on the site of what is now a block of flats. His father, Harry Marshall Ward, distinguished himself after graduating from Cambridge by travelling to Ceylon to track down and control a fungal disease which was decimating the island's coffee plantations. He returned home to marry Selina (always known as Lina) Mary Kingdon, who was the eldest daughter of a wealthy Exeter businessman. The Kingdons counted Sir Thomas Bodley, the founder of the Bodleian Library in Oxford, among their ancestors; they could also claim kinship with Isambard Kingdon Brunel, the great Victorian civil engineer. For centuries the family had distinguished themselves at Oxford and Cambridge Universities, at the Bar, in public service and the Church.

At the time of Frank's birth his father was a lecturer in botany at Owen's College in Manchester. A year later he was appointed to the Chair of Botany at the Royal Indian Engineering College at Cooper's Hill above Runnymede on the Thames, and the family moved into a solid, comfortable if not particularly beautiful house called The Laurels at Englefield Green, between Egham and Windsor in Surrey, which in those days was quiet and rural, in a gentle rolling landscape of hills and woods. It was in this setting that Frank spent his most formative years in the constant companionship of his elder sister, Winifred, born a year before her brother. (A third child, Dorothy, born in 1886, died in infancy.)

Frank and Winifred were devoted to each other, and remained so throughout their lives, despite long separations linked only by letters, and punctuated by tearful farewells on docksides and at railway stations and by joyous reunions.

While their childhood had all the solid order that characterised middle-class Victorian homes, their father's austerity meant that they enjoyed few luxuries or comforts, and none of the dolls' houses and rocking-horses that furnished the nurseries of their wealthier friends. The only time they ever had a Christmas tree was when they stole one from a public park. As a small boy Frank decided upon a career as a cab driver, and he spent hours astride a curved-topped trunk manoeuvering his invisible horse and cab through imaginary traffic. Perhaps because of their lack of material possessions the two children created a world of their own which they inhabited as Jack and May, living in permanent bliss and harmony with each other, whilst facing all manner of horrors and evil from the outside world. Fear played a prominent part in these games, including the grisly ritual of visiting The Dark Place.

In an unpublished manuscript based on their childhood, Winifred wrote: 'The visits to the Dark Place entailed some truly horrific experiences; but Jack and May, true to their characters of brave and noble beings, did not flinch from them when the occasion arrived. They knew that they must go for their souls' good, and they faced up to the visits manfully whenever it seemed the right time had arrived.'

The make-believe ordeal involved plunging into a terrible underworld of their own creation, reached through an imaginary well where they were decapitated before dropping into a river of blood filled with other headless bodies. The difference for them was they at first remained aware of the sights and sounds about them, which they described vividly to one another before eventually feigning unconsciousness. It was all in the best tradition of Victorian Gothic literature. The children shared a bedroom, and this oft-repeated ritualistic game normally took place in the early hours of the morning as it was just growing light.

One aspect of their upbringing was particularly unconventional by the standards of the time: the matter of religion. Their father was an agnostic, while the majority of the Kingdon family had been entrenched in the Established Church for generations. Lina's mother, on the other hand, was a devout member of the Plymouth Brethren, while a Kingdon cousin, Father George, was a Jesuit priest. Winifred was baptised, presumably under pressure from the Kingdons, and before her mother was persuaded into agnosticism by her father, himself influenced by T. H. Huxley. The two children were brought up on a religious switchback — as Winifred put it:

> It was a good old mix-up. Heaven (up in the sky) and Hell (somewhere down below), angels, harps and wings from nurses; strict Sabbath-keeping and prayer-saying from Granna, and a certain amount of the same from governesses; and at the back of it all, Father's rationalistic and scientific outlook, with a wise word in season regarding some of their superstitions; and Mother's tendency to shy away from the subject altogether.

This confusing start affected K-W's attitude to religion throughout his life. Sometimes he would pour scorn on the Eastern faiths he encountered, or unmercifully chastise the work of the missionaries. He would then quite suddenly switch to a more favourable view. Religion was a dilemma he was never able to solve, and his disillusionment with the power of prayer seems to have come early in life. He would often tell people: 'I remember one summer afternoon when I was a small boy asking God most fervently for a pop-gun to frighten the birds with. After waiting what seemed to a little boy playing in a big country garden an epoch, I got half the apparatus — the cork part; but the rest of the outfit never came along at all.'

While his upbringing as far as religion was concerned was unorthodox, in most other respects Frank had a typical Victorian childhood. His father's scientific liberalism did not extend to discipline. When teaching his children their arithmetical tables every mistake was rewarded with a slap on the hand with the flat of a paperknife, while minor sins were punished by being stood in the corner. Winifred recalled:

Apart from whippings, which might be administered independently or, in cases of very flagrant misdemeanor, might accompany it, being sent to bed in the daytime was about the worse punishment that could be dealt out; and it was not uncommon in that household ... It put a full stop to everything and seemed quite cruel to a degree; and the delinquent's pillow was usually wet with tears before he or she settled down into a resigned making the best of it — occasionally by going to sleep, but more often by singing to himself or making up stories or games in his own mind until deliverance came.

Professor Ward was, however, different from many men of his time in that he took a genuine pleasure in his children. He taught them to build snowmen, and amused them by demonstrating simple chemistry experiments. He was an accomplished musician and composer, and put poems by Kipling and Edward Lear to music, singing them in a fine baritone voice. He taught his children the songs of Gilbert and Sullivan, and entertained them with instant verses. When he found that the biscuit box only contained ginger nuts, for instance, he declaimed:

> I do perceive — I thought I should —
> Some biscuits that are made from wood;
> A dark brown colour, like the oak!
> Now am I not a funny bloke?

Or when the family cat tried to steal food from the table, he rose to the occasion with:

> Kitty came — ah well I mind her,
> Licked a plate of mutton broth,
> And departing, left behind her
> Footprints on the table cloth!

Allowing for the fact that most childhood memories become coloured by what was good, rather than the bad, there is no doubt that Frank and Winifred were happy, high-spirited children. Frank in particular was imaginative, mischievous, perhaps even a little unruly. Like many Victorian children of his background and class,

Childhood and Youth

Frank was forbidden to associate with the village children. It was a rule he cheerfully broke. Although he was small, and rather cherubic-looking with his curling fair hair — dressed in his best suit of dark blue velvet, with lace at the collar and cuffs, he could easily have passed for Little Lord Fauntleroy — he would deliberately pick arguments with the local boys just for the fun of a good scrap. When he was four, he and his sister and mother spent the greater part of a year in France, living in a house rented by Granna near Bordeaux. They crossed the Channel by paddle steamer, went by train to the Garonne, and finally by river steamer to La Tresne. The children were sent to a convent school to learn French. Frank was expelled after a week. In his first book, *On the Road to Tibet*, a collection of articles he wrote for the *Shanghai Mercury*, he recalled the incident with a good deal of relish:

> I had a gay time, and they stood for it for a week while I periodically burst into the girls' schoolroom bringing caterpillars and other frivolous pets which I had found in the garden to the Mère Superieure, who didn't want them.
>
> But the climax was reached when I danced into the chapel during Mass, to fetch a little girl of whom I was very fond, to play hide-and-seek with me in the grounds.
>
> The Roman Catholics are a susceptible people, and my crime could not be condoned. I was expelled forthwith, and the Mère Superieure herself taking me back to my mother's carriage said, 'Il a le diable dans ses jambes' (He has the devil in his legs) with tears in her eyes, thinking of my enormities, and my mother was very much ashamed of me for quite a week.
>
> No jam for tea, and everything I wanted must be asked for in French. There were quite a number of things I had to go without — it was very distressing.
>
> But I didn't cry, though I liked the convent school and the little girl; in fact expulsion from my first school was always a source of hilarity to me.

Whether the expedition to France triggered his passion for travel can only be a matter for speculation, but a remark overheard when he was a very young boy struck a spark that was never to be extinguished. Throughout his childhood, botanists and other scientists returned from overseas travel frequently visited his father. He heard one, a Mr Fisher, just home from India, say, 'There are places up the Brahmaputra where no white man has ever been.' He had no idea what or where the Brahmaputra was, but the remark left a lasting impression, as did the resonant sound of the name of the great river.

In 1894, following in the Kingdon tradition, he was sent away at the age of nine to board at Colet Court, the preparatory school for St Paul's in London. The next year his father was appointed Professor of Botany at Cambridge University, and the family moved again. It was in Cambridge that he met and became friends with a boy who was to have a great influence upon him.

Two years younger than Frank, Kenneth Ward was the son of the Professor of Moral Philosophy at the University. He was good-looking, daring, ingenious and athletic. On his last day at Oundle, a school he heartily disliked, he climbed the 215-foot spire of Oundle Church and tied his handkerchief to the weather vane.

During their school holidays the two boys were inseparable, sharing a mutual delight in roaming the countryside and camping out, often quite deliberately subjecting themselves to discomfort. They normally scorned the use of tents or sleeping-bags, instead digging shallow holes in banks for shelter, sleeping under railway bridges, or in railway waggons parked in sidings. They called themselves The Giddy Oysters, and the more daunting the enterprise the more attractive it became. One of their most eventful journeys was from Cambridge to Oxford, towing their kit on a home-made trolley attached to both their bicycles. Halfway to Oxford the trolley had all but destroyed their bikes, and was itself hopelessly disabled. At Buckingham they left the bicycles to be repaired, and hired a boat to continue the journey by water, first on the small spur canal that linked Buckingham to the Grand Union Canal, and then by canal through Banbury to the Thames and

Childhood and Youth

Oxford. The weather was appalling – a bitter wind and drenching rain. In the journal they kept they wrote:

> Finding late in the evening after a long search a retired spot by the river, we moored our boat and set up the tent ... We then after some difficulty – the wood all about being damp – managed to light a huge fire to dry our things by. We retained only our shirts, and proceeded to scorch both ourselves and our clothes beside it, dancing round every now and then to keep warm. Badly needing supper (it was then late and dark), we made some tea, so called: awful greenish soupy stuff that was all the dingy little village shop nearby could supply; and tried to eat some food that they had to offer – and it was worse than the tea! That meal was a record stomach-surprise!
>
> We then got into the tent and huddled up into the clothes, still damp in places; but the scorching in others no doubt made up for that! We stuffed up some of the gaps in the tent with grass, but the bitter wind blew it over us.

On reaching Oxford they explored the colleges, and then returned by boat to Buckingham, where their bicycles had been repaired. After sending the trolley and luggage ahead by train, they set off, in the late afternoon, to cycle the seventy miles home to Cambridge. 'We had a ripping ride in the moonlight,' they recorded in the journal.

> It was a lovely crisp, starlight night, the moon shining clearly on all around ... the fields seemed slowly to revolve past us, merging into the mist and fading away into the grey night beyond. But now and again thick banks of vapour crept out of the general dimness, lurking here and there in the hollows, or rolling up the valleys and enveloping the meadows in their deep mantle. All this had a soothing effect as we peddled by, ever leaving fields behind and ever meeting fresh ones gliding phantom like out of the mist in front.

In the Christmas holidays, when there was no tramping or camping, they amused themselves with what they called, 'Nightlies' – escapades conducted in the early hours of the morning, which some people undoubtedly must have regarded as acts of hooliganism. The target would be a barn or shed, or a half-built house, which they would break into, and then light a fire and cook a meal. On the way home they would let off home-made fireworks in doorways, and occasionally stuff them through letter-boxes. If they were chased by a policeman, gamekeeper or a night-watchman, so much the better.

Although they both made their careers in the Far East – Kenneth Ward went to Burma in 1911 as Professor of Mathematics and Physics at Rangoon University, and while there undertook a number of journeys in the region – they never travelled together on an expedition. Perhaps they would have done so had Kenneth not died in 1927 at the age of thirty-nine. It would have been an interesting exercise with K-W, pragmatic by comparison, the dogged and determined collector, and Kenneth Ward, deeply influenced by the poet William Blake, a practising Christian Scientist with an idiosyncratic belief in the pure spirituality of the universe. They both shared a contempt for Western materialism, but K-W did not share the younger man's political passion. What is not in doubt is that they had a deep and genuine affection for one another.

After Kenneth Ward's death, Frank Kingdon-Ward wrote:

> ... Kenneth was the most lovable boy on a camping trip, or a paper chase, or a boating excursion, or any escapade, because he was always cheerful, helpful and good-natured; and his capacity for forgiving others and turning away anger was simply unlimited ... He was the heart and soul of all our undertakings and the originator of most of the fun.
>
> ... Boy or man, I do not think he ever did a mean thing ... Most boys have a touch of barbarism in their composition, and very many even a streak of cruelty. Kenneth had neither; he was human in thought and deed

Childhood and Youth

> ... And as a boy he had an unusual insight into social and political problems ... He was right-minded on the topics of the day. What was more, he had the rare moral courage — rare among boys — to maintain views which were jeered at and regarded as rank heresy by his companions.

Apart from Kenneth Ward, K-W does not appear to have made any lasting or close friends at school. Certainly he makes no mention of any from either Colet Court or St Paul's, but he appears to have enjoyed life at both schools. In fact, when he set off for his first term at his prep school he sat beside the cabman on the driving seat, waved his cap in the air and shouted 'Hooray!'

He went to his public school in January 1898. In those days it was at Deadman's Field, where the parishes of Fulham and Hammersmith meet. Housed in a monumental red-brick building designed by Alfred Waterhouse (the architect of London's Natural History Museum, and Strangeways Prison in Manchester) it was under the High Mastership of the equally monumental Frederick William Walker, who had taken over from the kindly but ineffectual Herbert Kynaston.

Under Kynaston's direction, of the twenty-five boys leaving each year, only six went to university — fewer than from any similar public school. But Walker, the son of an Irishman, came to the school with an enormous academic reputation, and a determination to raise the academic standards. It was his decision to move the school from its former location in the City of London to Hammersmith, then in the countryside, and it was through him that the school recovered its enviable academic reputation.

With his great shaggy beard, *pince-nez*, gown and mortar board, he must have seemed an awesome figure to the new boys arriving for their first term. A contemporary of Frank's at St Paul's, Compton Mackenzie, used Walker as the model for Dr Brownjohn in *Sinister Street*:

> Dr Brownjohn was to Michael the personification of majesty, dominion, ferocity and awe. He was huge of build, with a long grey beard to which adhered stale morsels of food and the acrid scent of strong cigars. His face was

> ploughed and fretted with indentations volcanic: scoriac torrents flowed from his eyes, his forehead was seared and cleft with frowning crevasses and wrinkled chasms.
>
> His ordinary clothes were stained with soup and rank with tobacco smoke, but over them all he wore a full and swishing gown of silk. When he spoke his voice rumbled in the titanic deeps of his body, or if he was angry, it burst forth in an appalling roar that shook the great hall.

That was certainly how he was seen by the boys, K-W among them. A more considered opinion is given by Michael McDonnell in his classic history of the school.

> The kindness of heart which he concealed under a stern exterior was totally free from the sentimentality which in some schoolmasters tends to make their pupils prigs.
>
> His unquestioned authority in the government of the school was due to the full measure of latitude, free from petty interference in non-essentials, which he allowed masters and boys alike ...
>
> No public schoolmaster of our day has more richly deserved the praise of Cowley to his old master at Westminster that 'he taught but boys but he made them men'.

Walker met exactly the requirements for a high master laid down by the school's sixteenth-century founder, the Dean of St Paul's, Dean Colet; he wanted 'an honeste and vertuose and lernyd man'.

It is clear that St Paul's and its High Master had a deep and lasting effect on Frank. In 1899 his first report was less than promising – 'French: poor; Divinity and English poor, especially Divinity; Mathematics: weak, careless; Drawing: very fair; General remarks: rather a curious boy. Apparently conceited and obstinate.' But after his unpromising start, Frank began to make his mark at the school. At the end of the Christmas term in 1901 he won the special chemistry prize for pupils under sixteen, won it again at midsummer, and in 1902 and 1903 he was awarded the same prize in the higher age group. When he left for Cambridge University

Childhood and Youth

in 1904 he was awarded a Science Exhibition worth £40 a year, and it was predicted that he would enjoy a brilliant career as an undergraduate, and a dazzling future as an academic. Tragically, his university career was cut short by the death of his father at the early age of fifty-two. An obsessive appetite for work had led to diabetes. Professor Ward left his family in very straitened circumstances, so severe indeed that Frank was unable to complete his full three years at Cambridge. We know very little of his time there, or how happy he was at his chosen college, Christ's.

A family friend, Professor Giles, who was Professor of Chinese at Cambridge, arranged an appointment for him as a junior master at the Shanghai Public School, and so at the end of 1906, halfway through his final year, he sat Tripos, gaining a second in Natural Sciences. A few months later, early on a grim day in March 1907, he sailed from Tilbury Docks in London for the East.

Despite his father's untimely death and the shortage of money that effectively ended any plans for an academic career, K-W (as I shall refer to him from now on) never lost his love of scholarship, or the 'serious habit of mind' he learned from his public school. His most enduring characteristic was always to do the job he set himself extremely well.

2

First Steps into the Wilderness

WHEN K-W sailed out of Tilbury Docks he was travelling to a China torn apart by uprisings and destabilised by the final convulsions of a corrupt and moribund dynasty. For the Chinese themselves there was nothing particularly out of the ordinary about this state of affairs. Since the founding by the Chin Dynasty of the First Empire in 221 BC, eleven dynasties had been baptised and buried in blood. Again and again huge regions had become virtually autonomous under the control of a war lord seeking a power base from which to attempt the launch of a new dynasty.

Despite invasions, uprisings, civil wars, and attempts by Western nations to establish diplomatic and trading bases in the country, China was able to remain aloof from the 'barbarians' beyond her boundaries. She had defied invasion simply by absorbing the invaders.

In 1793 Britain sent a friendly mission to China under the leadership of Lord Macartney. It achieved very little. But with the industrialisation of the West, China became an irresistible goal to the Western powers, both as a market and a source of raw materials and of manufactured goods for the home market.

Britain, France, Russia and Portugal all managed to gain a foothold. Japan also had her eyes on China, but her ambitions were more territorial than commercial. All these nations were quick to take advantage of the troubled state of the Empire – the Opium

Wars, the Taiping Rebellion and the Boxer Uprising. Each upheaval chipped away at the old order. The erosion gathered pace in the latter half of the nineteenth century when each of these countries by one means or another tried to secure a share in China's riches.

These powers naturally took advantage of the internal dissent stirred up by the emergence of a new Chinese middle-class intelligentsia, with its republican sympathies, personified by Sun Yat-sen. He was the son of poor peasant parents who eked out a bare living in south-west Kwangtung. He would almost certainly have lived out his life in obscurity and poverty but for his elder brother, who had emigrated to Hawaii, made money and was able to pay for his young brother's education, both in secondary school, and later in medical school in Hong Kong.

As a student Sun Yat-sen was quickly infected with Western revolutionary ideas. When he went to Tokyo to continue his studies he established the United League among the large Chinese student population in Japan. The League planned and plotted the overthrow of the Manchu Dynasty, and the corrupt and largely moribund Ch'ing administration. Following the disastrous Sino-French war in 1895 he decided to strike, with an uprising in Canton. It failed; the first of many failures.

By the time K-W arrived in China it was only seven years since the historic alliance of Britain, Germany, Russia, France, America, Italy and Austria had fought and subdued the Boxer uprising, and only three years before Colonel Francis Younghusband was to lead a military force into the holy and forbidden city of Lhasa in Tibet; an enterprise which helped to undermine the Chinese influence in that country. In 1908, while K-W was still a schoolmaster in Shanghai, the scheming and devious Empress Dowager, Tz'u-hsi, died. She was succeeded by Pu Yi, a baby of two and a half, and the country fell under the inept regency of Tsai-feng. A kind of sham democracy was set up with provincial assemblies in 1909, and a National Assembly in 1910.

The Shanghai that K-W got to know was the intellectual heart of the revolutionary movement in China, and the centre of revolutionary printing. It was also the most westernised city in that vast country; a city built by merchants for merchants, owing more architecturally to the City of London than to the graceful, peaceful sweeps and curves of traditional Chinese buildings. The Bund, with its massive office blocks, banks, hotels and restaurants could be found in any European centre of commerce and industry. At night the department stores in the Nanking Road were brilliantly lit to display furniture and fashion comfortingly like London, Paris or Berlin. And in the European residential sector of this remarkable trading city, the inhabitants fastidiously maintained a strict, almost ritualistic western way of life, as though in terrible fear that they might become infected with some incurable oriental virus and 'go native' before returning home to enjoy the fruits of their exile. They devoted their leisure time to yachting, tennis, shooting, entertaining themselves in clubs, at dances, with theatricals, concerts, international sports fixtures, racing and, rather oddly, volunteering for service with the city's fire brigade.

K-W was there to help shore up this expatriate way of life by teaching the sons of wealthy businessmen, both European and Chinese, in a transplanted replica of an English public school. Not for the Shanghai Public School was the learning of the Hundred Schools of Philosophy, or the analects of Confucius, or the teachings of Mencius, or the principles of Taoism, or of the Mohists and Legalists which individually and collectively had fashioned the mind of China. It was the educational philosophy of Eton and Arnold's Rugby that prevailed.

K-W did not enjoy schoolmastering, and it was a mercifully short episode which he scarcely ever mentioned in later life. His sights were always set on the mountains and forests of the East, such as those he had read about in books like Sir Joseph Hooker's *Himalayan Journal*. He had a small foretaste of what was to come during the voyage to China. When the ship, the *Nore*, stopped at Singapore he went ashore and spent the night in the open by the Bukit Timah road. The smell of the soil, the tropical air, and the vast star-filled sky of the East bewitched him. During the school holidays he escaped from Shanghai to the Dutch East Indies to explore Java and Borneo. It was in Java that he became fascinated by volcanoes, but fortunately without the fatal results that befell the great Scottish plant hunter, David Douglas, who while exploring volcanoes in Hawaii fell into a pit that had been dug to trap wild bulls. Its enraged occupant trampled him to death.

One crater K-W climbed into was that of the 10,000-foot volcanic mountain, Gedeh. Recalling the experience many years later, he wrote: 'I well remember how the vast empty sulphur-coated crater itself, with its smoking holes, gave me the feeling of being down in the very bowels of the earth, or on the surface of the moon. The complete absence of life after the luxuriant tanglewood forest on the outer slope was striking.'

Escape from the monotony of the classroom came after he had been in his teaching post for less than two years. Through the recommendation of an old family friend, Oldfield Thomas, the Keeper of Zoology at the Natural History Museum in London, he was invited to join an American zoological expedition, which was to travel six hundred miles up the Yangtse to Wuhan, and then cross to Tibet. His role was to help with the day-to-day running of the expedition and assist with the collecting. The invitation was irresistible, and his school granted him leave of absence. Because of the backing given to it by the Duke of Bedford, it became known as the Bedford Expedition. Its purpose was to collect animals, and K-W discovered two voles new to science, *Microtus wardii* and *M. custos*, and one shrew, *Sorex wardii*.

He did try to encounter something rather larger than a shrew. It was a leopard which, an old peasant farmer said, was stalking his pig. At the time the expedition was travelling in the country between the Yellow River and the Yangtse.

Writing in the appendix of a book published in Shanghai in 1911 – *Wild Life in China, or Chats on Chinese Birds & Beasts* by George Lanning, an ex-principal of the Shanghai Public School, K-W recalled:

> I agreed to take a rifle and lie in wait, for it was a magnificent moonlight night, though bitterly cold. I lay in ambush, guarding the watched pig and awaiting the proud moment when I should shoot a leopard, till 2 a.m., by which time I was stiff with cold, for there was something like 15 degrees of frost out in the open; but no leopard came. So I turned

back and went to bed fervently blessing the old man, leopards in general, and pigs in particular. I have not shot a leopard yet.

Indeed, he never did.

He also managed to make a small collection of plants, which he sent back to the School of Botany in Cambridge, though the expedition was not at all significant in botanical terms. Most important for him, the enterprise was an apprenticeship in exploration, although temperamentally he was not suited to travelling and working in the large group he now found himself in. He was something of a loner and this of course had its drawbacks as well as its rewards, as he discovered during that first expedition. He had wandered off the planned route and lost contact with the main party. Completely disorientated, he attempted to get directions from some Tibetans he met, but was quite unable to make himself understood, and they for their part exhibited signs of hostility. His only protection was a loaded and cocked Colt revolver, which he kept in his pocket. After fruitlessly searching for the route that would reunite him with the main party he found himself wandering in a desolate place in the gathering dusk where eight kites were tearing at the naked corpse of a baby cast on to the barren hillside in the traditional Tibetan sky burial. It was a sight which moved him greatly. He was never able to forget 'the eyes widely open, glazed in death, the arms outstretched and slightly raised as though to clasp a mother's neck, the lips parted as though to cry out and thus transfixed with the cry still unuttered'. It took him two days and nights, most of the time spent in driving snow, and with only a hunk of bread for food, to find and rejoin the main party. At night he sheltered under overhanging rocks 'cramped into cat-like positions for the sake of keeping out the wind and retaining some warmth'.

He did rejoin his men safely, and would have reason to be grateful for the lessons he learned during the Bedford Expedition: how to deal with a crisis without panicking, as well as the art of observation and description and, perhaps most important of all, that of dealing with the indigenous peoples of the countries he travelled in. To his distress he found that the behaviour of many European travellers towards the local people was wholly distasteful. He made his disapproval very clear in *On the Road to Tibet* when he wrote:

> Men have openly boasted that they have travelled half across the country without disbursing a single cash, and others have earned an ugly reputation for emphasising their orders by the display of, or theatrical suggestion of, physical violence.
>
> In dealing with Chinese as in dealing with any other uneducated people, especially Asiatics, it is sometimes necessary to employ force if you mean to have your own way; but there are ways and means and degrees of applying it, differing in men and beasts.
>
> Certain Americans, Germans, Frenchmen, even Britons on occasion, have left an indelible and despicable impression amongst the natives.

K-W had heard of one traveller (he never named him) who had left a trail of resentment across China because of his behaviour. He was disgusted by one of the Bedford Expedition for displaying 'far more brutish stupidity than the ignorant, kindly peasants amongst whom we travelled'.

During his long career travelling in wild places, however, he did not hesitate to use force on the rare occasions when he considered it necessary. Being slight of build he usually had to take the initiative on these occasions, but he certainly never employed violence merely to assert his superiority. In 1913 he turned on his own escort of Chinese soldiers when they began to abuse and bully his Tibetan porters. One soldier pulled a knife on him. 'After that we had a fight, and finally the warrior tried to brain me with a brick. However in the end peace reigned.'

In 1928 while travelling in the Mishmi country he was lodging in a house when early one morning three young men burst into his room armed with knives. They objected to him being in their village. He dressed them down as though they were naughty schoolboys, took two of them by the arm and pushed them out of the room. With the understatement which was so typical of him, he recalled: 'They went quietly, but never stopped talking. My heart was beating rather fast, I noticed, but I carried the thing off. One of them might easily have drawn his knife and cut me down; and the last thing I wanted was a fracas when I was so near the end of my task.'

A few years later during the 1933 Tibetan expedition he got into a scrape which was even more dangerous, although he managed to make it seem almost comic. He was staying in a large house, a *dzong*, where his *sirdar*, Tsumbi, got extremely drunk and aggressive, and started a fight. K-W tried to break it up, only to attract Tsumbi's drunken rage. 'I seized his thumbs, ju-jitsu fashion; in a moment he was on the floor. He uttered no sound when I threatened to break both his thumbs if he showed the slightest sign of violence. He was quiet enough and I let him get up.'

But it did not end at that. After more drinking Tsumbi started another fight with one of the porters. Again K-W intervened, but to no avail. 'I gave him one more chance, then as he advanced I hit him between the eyes. The blow checked, but did not stop him; recovering himself, he aimed a savage but quite wild blow at me, grunting furiously.' At this point Kele, Tsumbi's brother, who was in floods of tears, grabbed K-W round the knees.

> ... he held me tight, and I realised that he had thrown his weight into the scale on behalf of his brother; he was drunk too. I then hit Tsumbi hard, and he went down. At the same time Kele threw me, and of course I fell straight into the arms of Tsumbi, who was now like a maniac. I saw murder in his eyes, but I grasped his thumbs again and he was helpless. Tashi [a loyal servant] now pulled Kele aside, and after a minute I sprang up.

Somewhat subdued, the brothers were padlocked into the caretaker's room, where they proceeded to try to batter down the door. Meanwhile K-W barricaded himself into his room and waited with a three-foot cudgel in his hand. Silence fell over the house, due largely to the fact that the brothers had been locked in the room containing the household stock of crude Tibetan rice spirit.

When they came too in the morning, however, they did succeed in smashing down the door. K-W was convinced that Tsumbi would attempt to murder him. The servant had lost face and would seek revenge. Despite this K-W decided to confront him.

> ... before I was half way, a figure appeared coming slowly down the gallery towards me. It was some seconds before I recognised Tsumbi, and I waited while he slowly and deliberately approached. He was a dreadful sight. He had a black eye, swollen lip and a cut forehead; a smear of blood striped one side of his inflamed face. His clothes were torn and dusty, his thumbs which I had wrenched back were swollen and discoloured.
>
> His pigtail had come down and his tangled hair hung over his forehead. But the look in his bloodshot eyes was defiant, insolent and alarming. I had purposely brought no weapon with me, and seeing the pitiable state Tsumbi was in I decided, while relaxing none of my vigilance, to give him a chance of making peace.

Tsumbi did in fact break down and beg forgiveness, so finally peace was restored.

K-W was certainly not a violent or bullying man by character. His widow, Jean, who accompanied him on all his expeditions after the Second World War, remembers him once disarming a man who was threatening him, but without having to use force. She never once witnessed him using physical violence against anyone. Nor does she believe that he was a naturally fearless man. 'He was no unthinking, unimaginative person who just didn't have any fear of anything. He was not at all like that. He had the normal fears that people do have, and overcame them,' she recalled.

He certainly was a product of the Empire, and shared the view of many of his contemporaries that the British writ ran good anywhere. Where he differed from many travellers of his time was that he did not take the line that it was his right, as an Englishman, to patronise, and that it was the duty of the natives to obey. He respected and observed local customs and laws even if he did not agree with them, or found them crude and barbaric.

First Steps into the Wilderness

In 1933, for example, a Mishmi mail runner sent to collect mail and a box of silver ingots was found murdered, with massive knife wounds in his head and chest. The mail was recovered, but the silver was stolen. K-W sought the aid of the Governor of Zayul at Rima, in the area where the crime was committed. Three men were arrested as suspects, and after some preliminary questioning were flogged with rawhide whips to extract a confession. K-W was obliged to watch, indeed he was asked if he had any objections, and when told it was the custom of the country, said: 'I do not wish to interfere in this matter. Please act according to the custom of the country.' However, he declined to award a number of lashes when invited to do so, and was able to intercede on behalf of a lama, whom he felt sure was telling the truth.

It was not only his ability to become part of the country through which he travelled that made K-W an outstanding explorer, but also his genius for absorbing and describing it. He had an easy, vivid writing style. As an apprentice traveller on the Bedford Expedition he demonstrated the keenness of eye and mind that remained with him throughout his long career. Finding himself for the first time in truly unspoiled wild country, he wrote:

> For two entire days the trail over the mountains led through a gorge of exquisite beauty, the cliffs, covered with pine and bamboo, rising abruptly two or three thousand feet above the ice-choked stream, making a grand play of colours in the winter sunshine.
>
> At one time alongside the frothing torrent, at another giving precarious foothold amongst the bush-clad precipices which yet towered far above, the trail wheeled sharply round bend after bend as it followed the sinuous curves of the river, affording endless views of matchless beauty. Beyond the projecting cliff which dipped boldly into the racing green waters, framing the picture for a moment, one more peep into the kaleidoscopic scenery always awaited us. A tiny white temple nestling amongst the dark pines which clothed a tongue of land; a glowing

> ochre scarp, several hundred feet high, crowned by waves of feathery bamboo; a sweep of firs hidden in a dark ravine, powdered with lingering snow that the dull warmth of winter had laid no hand on, all softened and blended in the mellow sunlight, and vignetted against the streak of twilit sky as night came on, ever changing in colour and in form.

That first trip implanted in him the most enduring love of rivers, which had first been stirred as a boy when he rowed for St Paul's on the Thames. Of his first experience of the great Yangtse, he wrote:

> High ranges of mountains on every hand, and the air was filled with the roar of cataracts and spinning whirlpools; the river sometimes broad and agitated, sometimes slinking along quietly, still and deep, between closely investing hills, seemed to be gathering itself for a spring. Whether calm or sedate or frolicking madly over the rocks, it was plainly working itself up for something better than we had seen before.
>
> Sunset behind us over this wonderful river was always a blaze of colour, but we were speeding away from it even when it beckoned us back to the wild west; night with its moonlit sky filled the gorge with rippling silver which danced over the gurgling water. So we drifted on past cities and mountains, past island temples and lovely pagodas, past orange groves and many a copse and spinney, dashing through rapids, whirling helplessly round in the big racing eddies which burst around and beneath us with the roar of approaching thunder; till early on 7th September we reached K'weichow-fu on the borders of Szechwan and Hupei, and saw before us the cliffs of Bellows Gorge towering to the sky, and the dark slit into which the river plunged.
>
> Suddenly the sun was hidden behind the mountain peaks, a blast of cool air whistled out of the ragged rent in front of us, and we sped between vast cliffs rising over

three thousand feet on either hand; we seemed to be speeding to some frightful destruction, in the bowels of the earth.

There were no rapids now. The confined water boiled and writhed around us; huge whirls sprang into existence everywhere with the crash of the avalanche, and we were at the mercy of the river, broadside on, stern first, flying round and round, while the *loa-pan* stood at the tiller yelling at the four oarsmen who stood on the forward deck, rowing like men demented to keep the boat's head straight wherever possible.

Nor did life on the river escape his attention, such as the glorious day they met, head on, the governor of a rural city sailing downstream.

It was a regular river carnival. First came a police boat, gay with flags, manned by eight or ten policemen in gaudy red tunics, these amphibians having turned into bargees for the time being; then came junks with the great man's furniture sticking out of the windows, piled on the deck, or tied to the mast; junks with his wives — also sticking out of the windows; junks with his children, and junks with his accessory relatives down to the third or fourth generation.

Finally came the great man himself surrounded by several more police boats, followed by a hustle of small boats, like threshers following a whale, all eager to pick up anything which the procession dropped overboard.

In those pre-Maoist days the Chinese were at liberty to subscribe to any or all religions — Buddhism, Taoism, Confucianism, even Christianity. The tendency was to turn to whichever belief seemed to fit the needs or ambitions of the moment. It was not unusual for a Chinaman to make sacrifices or take part in the ceremonies of more than one religion; a practice which bolstered K-W's often cynical attitude towards Eastern religions. However, even he was unable to remain unmoved by the spectacle of

Buddhist ritual which he observed during the Bedford Expedition in a lamasery in Choni, a village in Western China. He found hundreds of sandals in heaps outside the building:

> ... taking off my own boots, I entered the temple. It was very nearly dark inside, and the sombre gloom was accentuated by dozens of tiny butter lamps which burned like stars before the shrine of Buddha at the far end of the temple, and by a single shaft of sunlight which shot down from the centre of the lofty roof, touching with colour the strings of dingy prayer flags stretched across from side to side, and lighting up the massive pillars of the colonnade. Seated cross-legged on strips of carpet, their soiled brown gowns pulled loosely over them, eight long rows of dusty figures stretched up to where the bloated living Buddhas, resplendent in red cloaks edged with faded yellow, sat on their thrones beneath the shadow of the golden image; and through the mystic twilight the chanting of two hundred priests came like some wild voice out of the night, rising and falling on the wind and echoing through the mountains.
>
> Then all was still again; only one of the Buddhas was speaking in a hollow sepulchral voice, though the sound floated down easily enough; and again came the responses chanted by two hundred deep bass voices.
>
> All of a sudden the chanting ceased, and the band crashed out, reverberating strangely through the wooden temple, the ringing jingle of the cymbals, the roll of drums, the mournful wailing of the conch-shells, like some dumb animal screaming with pain; a mad roar of sounds, contrasting oddly with the dirge-like praying of the priests. This noise too ceased as suddenly as it had begun, and again the dark temple was filled with the swinging sound of men praying.
>
> Presently someone struck a stone gong three times, and as the dull echo died away the chanting ceased.
>
> Instantly a dozen priests entered bearing large baskets of wine and plates full of thin butter slabs, which they

First Steps into the Wilderness

handed along the rows of dry-mouthed supplicants, each of whom produced a wooden vessel from within the folds of his gown. The interval for refreshment had arrived. It was all inexpressibly weird in that gloomy temple, with its dead and living gods, its flickering lights and shifting shadows, and its bursts of prayer and wild music, as I stood, the cynosure of every eye, listening with bated breath to the blare of the conch-shells and the roll of drums, drinking in to the full the strange fascination of a pagan service.

The Bedford Expedition and his first solo journey taught him more about China, the real China, than most Europeans then living in Shanghai or Peking would ever know in a lifetime in the country. It was scarcely surprising that within a few weeks of returning to his job at the Shanghai Public School he was finding life almost unbearably irksome; the dry, chalky atmosphere of the classroom was no substitute for the gorges of the Yangtse or the immutable pageantry of a Buddhist temple ceremony.

How was he to get travelling again? He certainly could not finance an expedition out of a junior schoolmaster's stipend. There was little chance of getting any backing from the rich businessmen of Shanghai, who were more interested in exploitation than exploration. His experience with the Bedford Expedition, while invaluable, hardly allowed him to lay claim to being a seasoned explorer. But if he was to join the ranks of men like Carter and Andrews he had to find a backer, or devise some other way of financing an expedition.

He did receive a cautious approach from the United States Department of Agriculture Bureau of Plant Industry. David Fairchild, the Agricultural Explorer in Charge, had had him recommended as a potential explorer by the School of Agriculture in Cambridge, and had obtained his Shanghai address from his mother. In November 1910 he wrote from Washington 'to find out from you first hand what your ambitions are in the way of exploration'. Fairchild went on to say: 'I know of course that you stood high in your science studies at Cambridge but I do not know yet whether your training has been so much that of a field man as a laboratory scientist. The men we have sent out in search of plants for introduction have always been more or less field men, especially those who did the pioneer work covering a large territory.'

The US Department of Agriculture was keen to collect and study a number of Chinese economic plants, including ginger, persimmon, and the wood oil tree (*Aleurites fordii*). Fairchild explained that sooner or later these and many other plants would 'need a careful and thorough investigation by trained botanists who can see not only the botanical side of the problem but its practical bearing as well'. But then, with crushing frankness, he added: 'The fact that you are not familiar with the climate and agricultural development of this country would of course be a handicap, for you would inevitably introduce things which we would have great difficulty in finding a place for in our agriculture.'

It was not an encouraging approach, although Fairchild did say that he hoped that the Chief of the Bureau, Doctor B. T. Galloway, then in Japan recovering from a serious nervous breakdown, might call on him if he went to Shanghai. Fairchild did in fact keep in contact with K-W and later commissioned him to do some collecting for his Department. But in 1910 it was clear to K-W that his career was not going to be launched with American backing, and at the time he gloomily forecast for himself a 'humdrum life with every prospect of becoming a quiet and respectable citizen of Shanghai'. Then, out of the blue, came a letter from England that transformed his life. It was from A. K. Bulley, and was to start him on his distinguished career as a professional plant hunter.

At that time, fired by his experiences on the Bedford Expedition, his mind was set on becoming an explorer, pure and simple: he wanted to discover the undiscovered, question established beliefs, advance theories based on his own work in the field, and, above all, go where no other explorer had been before. In fact he never ceased to see himself as a geographer, which was why he was later subject to criticism from academic botanists that his botanical work sometimes lacked scientific thoroughness.

First Steps into the Wilderness

One of his closest friends and his executor, Sir George Taylor, former Director of Kew Gardens, found that his herbarium specimens tended to be fragmentary, and that he occasionally confused a natural hybrid with a specie. Sir George recalls catching him out over the meconopsis which K-W introduced as the Ivory Poppy, and a specie in its own right. 'I said it was a cross between *M. simplicifolia* and *M. integrifolia*. Later when I went to his camp site at Temo-la, where he found the poppy, I found the parents. I was right and he was wrong.' But what his fellow botanists could not fault him on was his unerring eye for a good garden plant, and that was what Bulley wanted.

Arthur Kilpin Bulley was a wealthy Cheshire cotton merchant with a passion for plants and gardens. A brilliant amateur naturalist, he had a particular interest in introducing exotic hardy plants to Britain. To this end he established his own seed firm, Bees. To start with he got his seeds from missionaries and enthusiastic amateurs, but the quality of the material was poor, and many of the plants raised from the seeds were hardly, if at all, superior to common English weeds. He realised that what he needed was a trained botanist with a keen eye for first-class garden plants. From correspondence with people such as Augustine Henry, who worked for the Chinese Customs, but devoted much time to botany, he knew that the interior of China was rich in marvellous plants just waiting to be collected. A close friend, Bayley Balfour, the Director of the Royal Botanic Gardens in Edinburgh, recommended George Forrest.

Forrest made two trips for Bulley, the first in 1904. Both were spectacularly successful, but then Forrest was, according to Bulley, 'poached' by the distinguished gardener and plantsman, J. C. Williams of Caerhays in Cornwall. This left Bulley without a man in China. For a while he had to content himself with taking shares in expeditions financed by other enthusiasts like himself. However, Bees was now becoming famous, and its 1911–12 catalogue was able to offer 'New and rare hardy plants, including our own introductions'. At the same time Bulley was well advanced with the development of his famous garden, Ness, at Neston in Cheshire, which his widow left to the University of Liverpool. He could no longer rely on seed shares, but must have his own man in the field.

Once again he consulted Bayley Balfour, who remembered that the son of his old friend, Marshall Ward, was in Shanghai. He suggested that the young schoolmaster should be given the commission. When he received Bulley's letter, K-W said that he did not even stop to consider the consequences before deciding to accept the job, which was to collect hardy alpine plants from the mountains of Yunnan and along the wild Tibetan marches. This time there was no question of a temporary suspension of his teaching contract. He walked out of the classroom for the last time, and did so with a light heart.

It was only when he paused to reflect on what he had done that he began to question whether he had made the right decision. His sister Winifred said that as a boy and young man he was never over-confident, either of his ability or his knowledge. Now he was faced with going alone to a remote part of China, which he had only visited once before and then as part of a large expedition for which he had no real responsibility. He also had to withstand the well-meaning criticisms of friends and colleagues who saw him throwing away a secure future for what seemed to be little more than some madcap adventure. Curiously, he was himself in later life rather resentful that he was not paid a pension for his brief period of schoolmastering, despite the fact that it was he who broke the contract. But however great his fears and doubts, he knew with certainty that his future lay in the wild places, not the safe sinecures. So on 31 January 1911 he set off alone for Yunnan.

3

The Making of a Plant Hunter

KINGDON-WARD's first solo expedition has been beautifully recorded in his book *The Land of the Blue Poppy*, arguably the best of his twenty-five books. It has a vigour, freshness and excitement which mirrored his own response to the great adventure. With the Bedford Expedition he had been an apprentice, but as Bees' commissioned collector he was entirely responsible for the 1911 expedition, and his whole future depended on its outcome. It was as though he was sitting his finals with a worrying amount of study still to be completed.

Instead of travelling overland he approached Yunnan through Burma, sailing from Shanghai to Rangoon, and then making his way by train to Bhamo. The first part of the journey had a holiday air about it. Knowing that he was embarking on his first solo expedition, his fellow passengers regarded him as something of a young hero. It was not until he left Bhamo that the reality of his situation dawned on him, and he felt a sense of desolation. He wrote:

> The initial stage out of Bhamo is only nine miles, and it was undoubtedly this fact alone which caused me to feel extremely lonely on the first evening of my journey. Arriving very early in the afternoon there was of course nothing to do but to take out a gun and look round for game, but, do what I would, there was no getting away

> from the sense of utter desolation which seemed to crush me. Even the mild excitement of putting up a barking deer among the reeds of the river failed to alleviate the depression and after dinner I was only too glad to crawl into bed and, weary in spirit, court oblivion in sleep. Never again did the sense of paralysing isolation come so vividly upon me as on the first night when all the trials that awaited me seemed to take shape and rise in arms to mock my ignorance and feebleness.

While it is probably true that this was the first and last time that he seriously doubted his ability to fulfill a commission to find and collect new plants, it was not true that he was never again seized by a devastating sense of loneliness. That and bouts of black depression struck him many times during his career, but they tended to be dissipated by the splendour and beauty of his surroundings, the glory of the plants, and the thrill of the hunt. On this first solo journey his depression and loneliness evaporated at the sight of blazing scarlet rhododendrons and masses of pink camellias. They made him 'almost oblivious of everything else'. He found himself walking among 'sheets of mauve primulas, dog-roses yellow and white, pale blue irises, and other delightful flowers'.

He never really needed to doubt his ability, returning as he did from the expedition with a collection of two hundred plants, twenty-two of them new species. They included *Silene rosaeflora*, five new saxifrages, among them *S. wardii*. There were two new sausureas, a cremanthodium, and a meconopsis named after him, *M. wardii*. Other plants from the expedition which were given his name were *Gentiana wardii*, *Listera wardii*, and *Androsace wardii*. He also introduced *Primula vernicosa*, the first of what was to be a long list of fine new primulas. What were missing from his collection this time were rhododendrons, plants for which he was to become famous.

Harvesting the seed was not a simple matter of picking capsules and putting them into packets. In the first place the seeding plants, which had been seen in flower months earlier, had to be re-located. If a particular species was very abundant then it was not necessary to mark it; for others, directions would have been written down, such as 'forty paces from the magnolia on such a track in such-and-such a place'. In certain cases a plant would be marked with a coloured ribbon or a piece of yarn. In fact K-W had the most astonishing memory for the location of the plants he discovered. Retracing his steps for the seed harvest, he would go straight to them. Even when the terrain was covered by snow, he could dig and find plants and gather the capsules being held in frozen state for the spring germination.

Plant hunting was not a matter of collecting seeds. The herbarium collection had to be completed. Every species had to include a flowering specimen and the same plant in fruit. And if you were supplying more than one museum or botanical garden then you had to collect more than one sample.

At the end of a day in the field, K-W still had hours of work ahead of him when he returned to his base. All the herbarium specimens had to be sorted through, and the best selected for pressing so that they would look good on the sheets. Each collected plant had to be numbered and listed. Every plant hunter starts his collecting career with number one set against his name or initials. When he died K-W had over 23,000 plant numbers to his credit.

It did not end by putting plants between the thick, absorbent sheets of paper in a plant press. The presses had to be opened every twenty-four hours to get air to the specimens until they were bone dry. When the weather was very wet, which it naturally was during the rainy season, even the thoroughly dried specimens had to be examined regularly to see if they were being attacked by mould. If they were, it had to be brushed off with a very soft shaving brush.

Seeds required equally careful treatment. Plenty of pockets, tins, and cheap sponge bags were needed in the field to hold the harvest. Back at base the seeds were laid out on waterproof sheets to dry, and were turned every two hours or so. When they were quite dry they were winnowed by blowing away the chaff before being packaged with an identification slip, or perhaps a leaf or late flower. Biscuit tins made ideal containers for packets of seeds, and when stores were being ordered before an expedition, the usefulness of containers was an important consideration. The tins were then packed into empty boxes which had been used for carrying supplies. If the weather was wet the packets would be spread out in the open whenever there was some sunshine.

Although he normally stuck to the tried and tested methods, on at least one occasion K-W did experiment with seed packaging. In 1921 he tried a technique of packing seeds in a bath of carbon dioxide. The seeds were put in an airtight tin and a soda sparklet was used to introduce the gas, and then the container was sealed with solder. The seeds germinated well, but then so did the seeds packed in the conventional way sent with the same consignment, so the extra trouble was not worthwhile.

Bulbs, tubers and orchids with their pseudo-bulbs were packed in moss or soil in locally made woven baskets or empty store boxes. After the Second World War with the improvement in air services it would become a great deal easier to ship living plants to backers.

While relatively happy with the demands of plant collecting, K-W was less prepared for the fact that local people in remote parts of China and Tibet endowed European travellers with almost magical medical skills, and K-W would have been the first to admit that he was a rudimentary doctor, to say the least. One of his first patients, as he made his way to the hunting grounds for Bees, was a muleteer with a badly cut foot. Before he could clean the wound he had to hack his way through hide-like skin. Whether the other men in his caravan were impressed by his performance is not known. What really attracted them was the sight of his medical chest. 'All the muleteers discovered there was something wrong with them,' he recorded, 'not because they wanted to go back [he had sent the wounded man home to recover] but because they wanted to sample my tabloids.'

What really made him angry was when his best endeavours were rendered useless by what he saw as a deadly mixture of ignorance, superstition and a slavish adherence to religion. At Tsukou his Chinese landlord asked him to look at his youngest child, a baby of about a year old who was suffering from a bad chill and cough. K-W gave the little boy a few drops of chlorodyne, a then popular anodyne made from chloroform, morphia, *Cannabis sativa* and prussic acid, and some quinine tablets, and told his parents to wrap him up and keep him warm, and to feed him with hot milk. The following day when the boy was better his parents took him out into the bitter cold to be baptised. That night he died. 'Poor ignorant people,' he wrote, 'they were more anxious to save his soul than to preserve his body.'

He was not a demonstrative man, but he had a tenderness and compassion which manifested itself in the most practical way. Six months into the expedition his personal servant, Kin, fell ill. His chest, arms and legs were covered with hideous purple blotches, and he vomited constantly. Perhaps it is a measure of K-W's somewhat meagre medical skills that he diagnosed these alarming symptoms as 'a slight chill'. However, he dug into his own stores for brandy and a cholera belt. He made a mustard plaster for his chest and tried, unsuccessfully, to get him a riding pony, but he did keep him going with brandy, milk and soup from his own supplies.

Kin's special needs seriously diminished K-W's food stocks. 'I was now completely out of stores with the exception of some soups which I kept for Kin, who happily was feeling much better. I was still giving him brandy night and morning — though we were nearly out of that too — and feeding him chiefly on lightly boiled eggs.' K-W got by on *tsamba*, a flour made from parched barley, eggs, brick tea, fungi and bamboo shoots.

Apart from having to contend with illness, short rations, and the natural hazards of rockfalls, landslips, and freak weather, in 1911 a lone European needed courage and a cool head to travel in the area, for both Tibet and China were in crisis. The historic loathing of the Tibetans for the Chinese manifested itself in Lhasa in that year with the murder of the Chinese Viceroy — a position second only to the Emperor — and over one hundred of his followers. The massacre brought savage retribution. Scores of Tibetan men, women and children were beheaded or beaten to death by Chinese soldiers. In July K-W received a message from a French missionary priest which told him: 'The English are in

Lhasa, the Chinese soldiers have capitulated. A British soldier has gone from Y'a-k'a-lo on a secret mission. The Chinese are furious and swear to exterminate every Englishman. I fear you will be killed before the end. You must leave A-tun-tzu at once.'

He took the warning seriously and decided to leave A-tun-tzu for Batang. It was on the road to Lhasa, and he knew he could join up with a number of European missionaries there. Batang was the town where only a few years earlier armed lamas had massacred Christian converts and Catholic missionaries, and where the plant hunter George Forrest had barely escaped with his life.

In fact the message K-W received proved to be a false alarm, but while he was in Batang he saw for himself evidence of the savagery of the Chinese soldiers when he met a minor Tibetan prince recovering from fifteen hundred blows from heavy wooden clubs received as punishment for misgovernment. Normally the victims of such barbaric treatment died.

Four months later when he was back in his collecting grounds further news of unrest reached him, including word of the fall of the Manchu Dynasty. Of more direct concern were rumours of an uprising at Batang, and of Chinese soldiers deserting the army and taking to the mountains. That could only mean one thing – gangs of renegades robbing and murdering throughout the countryside. Shortly afterwards he received confirmation that Yunnan was in a state of revolution, and that the capital, Tali, and the important town of T'engyueh were in the hands of the insurgents.

He did have a Chinese military escort, but it gave him little reason for comfort, since the ill-treated, ill-paid soldiers were little better than bandits. However, he managed to keep control of the situation with a robust style of discipline. When he caught one of the soldiers with a kettle he had stolen from a hut where they had spent the night he was so angry that he punched him in the face. The soldier tripped over his rifle and 'fell in a heap on the ground, and lay like a half-empty sack of corn, bleeding from a cut lip'. With his Colt automatic pistol in hand, K-W marched the man back to the hut and made him return the kettle.

Despite his precarious position, made worse by the fact that all Europeans were regarded with the greatest mistrust by the

Chinese, K-W managed to strike up friendships with some of the revolutionaries, most of whom seemed more interested in warlordism than reform. One such man was a Captain Li, who led a raggle-taggle band of deserters and tribesmen. K-W met them marching from the Mekong into the mountains.

> They formed nonetheless a striking picture, winding in single file up the narrow ravine, rather more than two hundred of them, though the majority of them were baggage coolies. There were Lissus and Minchias, sturdy little tribesmen with muscular chests and swarthy complexions, often ferocious of aspect, carrying rifles, most of them muzzle-loaders; men with extraordinary-looking guns of abbreviated length and immense calibre, after the pattern of a blunderbuss; others carrying scarlet banners and long trumpets; and half a dozen men, including my friend Captain Li, riding ponies.
> How strange they looked, these poorly clad tribesmen armed with *dah*, crossbow, and arrow-case of black bearskin, each carrying a load of no mean weight on his back by means of a strap passing over his forehead. Occasionally the trumpeters put the long trumpets to their lips and throwing back their heads, made the welkin ring; and on they hurried, banners fluttering till they wound out of sight on the narrow gorge.
> What did it all mean? What could it mean but that Li was inciting the tribesmen to rise and fight for – what? An independent Western China?

In fact what was behind the recent riots and uprisings was a row over the construction of railways to the remote provinces, particularly the Hankow-to-Canton and Hankow-to-Szechuan lines. The right-wing Ch'ing government wanted the Western powers to finance and build the railways, hoping that in return the foreign governments would prop up the ailing Ch'ing regime. The Republicans wanted the enterprise to be exclusively Chinese, and to see an end to foreign influence in China's affairs. Despite brutally suppressing the rioting which resulted, the government and the

Manchu Dynasty collapsed. The young Emperor, Pu Yi, abdicated on 12 February 1912, and the following day Sun Yat-sen, who had been declared President Designate of the Republic of China, resigned. The scholarly architect of Chinese republicanism had been out-manoeuvred by a powerful army leader, Yuan Shih-k'ai, who had been appointed provisional Vice-President.

It was evident that Yuan wished to become dictator of China, but in a desperate bid to create a constitutional parliament Sun Yat-sen and his followers agreed to support Yuan for the time being in exchange for the promise of parliamentary elections. These were held with Yuan leading the Chinputang Party, and Sun Yat-sen at the head of the Kuomintang.

The Kuomintang won the election. Yuan responded by ordering the murder of the Kuomintang premier-in-waiting, Sung Chiao-jen. Sun Yat-sen and his followers decided to fight fire with fire. They launched what was called the 'Second Revolution'. It was short-lived. Yuan had the army on his side, and the Republican leaders were forced to flee the country. Believing himself to be completely in control, Yuan proclaimed himself the Emperor of a new dynasty. But his reign as a Son of Heaven was brief. China had had enough of monarchies, the army turned on him and after four years he was forced to abandon his grandiose role. The humiliation killed him.

It is necessary to digress from K-W's first solo expedition into the complexities and violence of twentieth-century Chinese politics in order to illustrate the problems with which he had to cope. It was difficult enough to deal with the hazards of travelling among unpredictable tribes in dangerous, largely unexplored country, without the added dangers of revolution and warlordism. All his difficulties, both national and local, he handled with classical British phlegmatism. He was frequently faced with potentially dangerous disputes, such as a disagreement with some Miao muleteers during that first solitary journey. He walked into a room to find himself confronted by the men. Three had wooden benches raised above their heads as if they intended to brain him, while another was aiming what appeared to be a gun at his chest. K-W was unarmed and there was nothing for it but to brazen it out. He smiled cheerfully at them, and jokily talked them out of their angry mood. The 'gun' turned out to be a bamboo spear, none the less dangerous for that. But the episode taught him an important lesson that was to serve him well in the years ahead. 'After that I was more careful than ever in my dealings with tribesmen who were unaccustomed to Europeans,' he wrote.

It would be wrong however to create the impression that his journeys were dogged by crisis. More often than not K-W was treated as an honoured guest, and the local people, particularly the Tibetans, for whom he had a deep affection and admiration, would go out of their way to make him welcome and to entertain him. In *The Land of the Blue Poppy* he describes an evening when he was lying comfortably in a pine-log hut, writing up his diary for the day:

> ... in stalked three Tibetans, all of them over six feet high. Their coarse gowns were tied above their knees, the right shoulder thrust jauntily out exposing the deep muscular chest, and they were bootless. One of them carried a fiddle, consisting of a piece of snakeskin stretched over a bamboo tube with strings of yak hair, upon which he scraped vigorously with a yak-hair bow. There was little enough room, but my visitors soon lined up, stuck out their tongues at me in greeting, and began to dance, to and fro, up and down, twirling round, swaying rhythmically to the squeaky notes of the violin (there were only two notes on which to ring the changes), and singing in high-pitched raucous voices. Presently three women joined in, all tricked out in their best skirts and newest boots, with cloaks flung negligently over their shoulders.
>
> Thus they went through many of their national songs and dances, and in justice to my sex I must say the men danced with more skill and grace than did the women, though of course it is easier to dance heel and toe, barefooted like the men, than in the clumsy boots and skirts worn by the women.
>
> I can still picture the scene in that dim little smoke-

blackened room, the rain lashing down outside, and the roar of the river below us, while I lay back on my bed enjoying it hugely, all cares forgotten. Those great giants of men looked strangely weird in the flickering light of the blazing torches which flared up and burnt down alternately; the wail of the fiddle rose and fell, the voices blended, and broke, and ceased, and still they danced on, up and down, to and fro. They danced for two hours in all, and in return for a little present I gave them, would have willingly gone on till midnight had I not told Gan-ton I wanted to go to sleep.

Tremendous rainstorms that created instant torrents roaring down the hillsides; thunderstorms that sounded as though all the ordnance of the world had opened fire at once; hail that hissed on the alpine turf like spent bullets; snow storms; gales that filled the air with flying gravel; unheralded rock falls that buzzed like swarms of killer bees; petty theft; robber armies, revolution and political upheaval – none of these could deter him from the ecstatic pleasure he derived almost daily from the country he travelled through.

On reaching the Yangtse during his march to Batang after receiving the warning of danger from the French missionary priest, he recorded:

> ... never shall I forget the first view of that noble river as we climbed the last spur and looked northwards over the trees and towards Batang. The sun was down, and over the purple mountains great puffs of radiant cloud rested, scattering the dying light; for miles we could follow every twist of the valley, marked by a ribbon of flashing silver, which still had 3,000 miles to flow before it reached the ocean.

Some weeks later he was camped in a village called Samba-dhuka. In the early morning while his men were packing up ready for the day's trek he climbed to a high escarpment and looked down on to the Mekong.

Just here the river presented an extraordinary spectacle. For a quarter of a mile it flowed between fluted walls of limestone, not more than fifty feet apart and perhaps a hundred feet high, and looking down from the cliff on the red water writhing in this confined sword-cut below, gave one some idea of the irresistible power of the river.

The way it ground its way through solid rock to form great, bleak gorges filled him with awe, a kind of dread. He was completely captivated both by the River Mekong's harsh beauty, and its gentleness where it flowed quietly past villages:

> ... there was a strange fascination about its olive-green water in winter, its boiling red floods in summer, and the everlasting thunder of its rapids. And its peaceful little villages, some of them hidden away in the dips between the hills, others straggling over sloping alluvial fans or perched up in some ancient river terrace – all these oases break the depressing monotony of naked rock and ill-nourished vegetation, delighting the eye with the beauty of their verdure and the richness of their crops.

It was an unpitying country to travel through, and yet the exhaustion, the aches and pains at the end of the day, were easily forgotten with the majestic sunsets which would flood the landscape. Of one over the Mekong, he wrote:

> We were now some two thousand feet above the river and as the sun sank down behind the towering cliffs above us, the valley was filled with a glow of such wonderful colour that no description of mine can convey any idea of its spell. While all was dark and gloomy in the depth of the valley, the setting sun caught the tops of the mountains across the river, and one forgot their bare brown slopes under the waves of crimson light which they reflected. Gradually a deep blue shadow crept up out of the valley and wrapped the hills in slumber, while a soft clinging mist seemed to precipitate itself from the atmosphere and spread over the rice fields below. In the gloaming the crimson

died down to purple, the purple became violet, and still the glorious colours of the sunset played up and down the valley. Away to the south a few wisps of cloud caught the slanting rays of the sun, which flashed like the beams of a heliograph through a gap in the black wall of rock overhead, and diffused an orange glow into the deepening blue. Then a few stars shone out, and the ridge was clearly silhouetted against the eastern sky: night had come down like a curtain.

Throughout most of his first solo journey his closest companion was Ah-poh, a Tibetan mastiff which had adopted him. The animal might have been related to Montmorency, the idiosyncratic dog in *Three Men in a Boat*. It was obstinately independent, and the only concession it ever made to obedience was to come when K-W called. Enormous and hairy, it had a rugged sense of fun, specialising in stampeding caravans of laden mules, after which it would turn to its adopted master 'with a pleased "See what I've done" sort of expression'. Eventually K-W had to lead Ah-poh on a length of rope past pack mules, but even this precaution did not completely reassure the nervous animals.

Ah-poh disappeared in T'eng-yueh at the end of the journey. Clearly the animal was one of life's drifters. K-W was saddened by the loss. During his years of travelling he had a number of dogs, but none seemed to last very long. Years later, in a letter to a friend, Constance Levinson, he wrote: 'Keep a team of dogs, lavishing affection on them all. Lose them periodically, one by one. If one turns up, after being lost for a time, all the better. If it doesn't, write it off.' He wrote the letter just after hearing that a favourite dog had died.

The Chinese part of the expedition really came to an end in T'eng-yueh in western Yunnan. K-W arrived there at the end of 1911 after a three-week march during which he had not changed his clothes, bathed or combed his hair: '... my hair was long and unkempt, my face pinched and bearded ... my feet were sticking out of my boots, my riding breeches torn, my coat worn through at the elbows'. For all the discomforts and hardships, and the constant worry about the worsening political conditions in China, he felt an immense and justified sense of triumph. He knew that he had proved himself, had been fully blooded as an explorer, and had come through the experience successfully.

As well as the fine harvest of seeds for Bees — a number also went to the Royal Botanic Gardens in Edinburgh — which was the main purpose of the expedition, he also made detailed botanical and geological observations. From the latter he concluded that the great mountains of the Himalaya and Trans-Himalaya had been raised as the result of 'terrific lateral pressure, either acting simultaneously from east and west, or more probably from one side only, the other side being crushed against an unyielding barrier which, by preventing any actual movement of the mass so caught, has compelled it to ruckle up in parallel ridges as one might ruckle up a piece of cloth'.

The unyielding mass, he argued, was Western China. Although there had been volcanic activity in the area in the past, and earthquakes were still common, he believed that the great gorges were the result of the squeezing, that the same activity had determined the courses of the three great rivers — the Irrawaddy, Salween and Mekong — and that these rivers had dug the gorges deeper as a result of vast quantities of water from monsoon rains and snowmelt being forced into the relatively narrow fissures in the mountains. He noted the water-mark on the face of the gorge of the upper Salween in December, when the level of the river still had some way to fall, was thirty feet above the surface, and wrote: 'A rise of thirty feet in a river averaging sixty yards in breadth and flowing with a strong current implies a force which is almost beyond belief till one has seen it at work.'

He noted how the great ranges of mountains could alter the weather dramatically over comparatively small areas, producing local climates that changed suddenly from endless rain to almost desert conditions. His examination of the valleys that he was constantly entering and leaving convinced him that they had been carved out of the mountains by ancient glaciers.

A close study of the distribution of plants would, he believed, reveal much about the origins of the great rivers, and of the mountains through which they flowed. One theory he advanced was that if a plant flourishing in wet conditions was found struggling for survival in an arid area it could be argued that the struggling plants were survivors from a period when the dry region had enjoyed a wet climate.

> The sulphur-yellow *Meconopsis integrifolia*, for example, a plant which flourishes in a wet climate, occurs very sparsely on the Mekong–Yangtze ridge, being confined to a few favourable localities, whereas whole meadows of it are to be seen on the Mekong–Salween ridge. The plant is very common to the north and east, on the Szechuan mountains, whence it is reasonable to infer that it has come down these ridges, having since almost disappeared from the Mekong–Yangtze divide, though it is of course possible that we find there, not a remnant of a former extensively distributed species, but a plant which is trying to establish itself. Such cases suggest lines of investigation, but require a much wider acquaintance with facts to justify conclusions of any importance.

In fact the study of the distribution of plants has become a key factor in trying to determine the story of the formation of our planet. Seventy-five years ago K-W warned:

> The whole idea of attacking the geological problem from a botanical point of view, however, opens up such a wide field for investigation that it is useless to pursue it as an aimless speculation without marshalling an enormous array of facts in support of this or that contention. I have merely made certain suggestions based upon limited observations, the accuracy of which is in some cases unfortunately open to criticism, and I must leave it to others to say whether they are justified or the reverse.

After a brief reunion with his boyhood friend, Kenneth Ward, who had taken up his professorship at Rangoon University, K-W left Burma in January 1912 for England. Five years before he had set out from Tilbury as an obscure junior

The Making of a Plant Hunter

schoolmaster. He was returning a fully fledged explorer and plant collector.

He went straight from the ship to 23 Chesterton Road in Cambridge, where his mother and Winifred were living. His father's early death had left his mother in straitened circumstances and it was a smaller house than the one they had when he sailed for China. However, it was comfortable and ideal as a base. His mother had arranged for him to keep his last term at Christ's, which he had missed when he went to Shanghai.

The year 1912 was a full and happy one for K-W. He wrote *The Land of the Blue Poppy*, and was kept busy lecturing on his travels. Writing was an important part of K-W's life. He was a regular magazine contributor, and the vivid style of his books is evidence of his literary skill. The detailed observation in both his diaries and his letters reveals that he was planning a book from the first day of every expedition.

He wrote a considerable number of articles for publications like the *Gardeners' Chronicle* when he was actually out in the field, and the first drafts of books were started under canvas or in some flimsy hut, particularly when he was unable to search for plants because of the monsoon or trouble with native porters. He would then write them up for publication during his sojourns back in England.

Despite the effort he put into his literary work, it did not greatly reward him. He quite frequently changed his publishers, not because of disagreements, but simply in the search for someone to publish him, and too often the income his books brought in was woefully small. His deal with Cambridge University Press for *The Land of the Blue Poppy*, for instance, was half of the profits, but paid over only after all the expenses of publication had been covered. In 1923 he was paid an advance of only £30 for *Mystery Rivers of Tibet*, receiving a mere £26. 19s. 6d. after paying commission and expenses. For the second half of 1941 journalism and book royalties brought in only £15. 15s. 2d.; 1942 was better with £62. 18s. royalties, but the following year produced a measly £6. 2s. 8d., plus a four-guinea fee from the BBC. In 1944 writing yielded a total for the year of £216. 15s. 1d. Even allowing for its value against inflation over the last forty-five years, it was poor remuneration for all his hard work.

His lectures brought him to the attention of the Royal Geographical Society. Impressed by his imaginative approach in trying to probe the secrets of the regions he explored, the RGS that year elected him a Fellow, which made him the youngest man accorded that honour. In 1930, in recognition of all his achievements, they awarded him the Founder's Gold Medal, one of their highest accolades. E. H. M. Cox, who travelled with Reginald Farrer collecting plants in Upper Burma, was later to comment: '[K-W] is, and always has been, much more than a plant collector. He tells me that he is prouder of the Royal Medal of the Royal Geographical Society than of his three gold medals for horticulture. There is no doubt that pure exploration is his first love, but, alas, it cannot be commercialised and one cannot live by exploration alone.'

The RGS also indicated that it would be prepared to provide him with surveying equipment for the next expedition he was planning. Towards the end of the year the Society wanted to know exactly where he intended going, but the unrest in China made this an almost impossible question to answer.

On 23 November 1912 he wrote to the Secretary of the Society, Dr Scott Keltie:

> China is at present a closed country, at least so far as the interior is concerned, and I am not sure how far it is wise for me to say I hope to reach this or that place; moreover I may be making a boast which, with the forces arrayed against me, it may be impossible to fulfil.
>
> I do not of course know who is on the Council of the RGS, and it is quite possible that someone might feel it his duty, in another capacity, to completely put a stop to any ambitions in that direction, as far as it lay in his power. Still, you will know this better than I, and I am quite ready to let you know that I propose to go back to the region I was in last year, only more so, i.e. further in. I hope, if things turn out well, to get into unexplored as well as

unsurveyed country, but naturally there is a good chance of the whole thing failing on account of the distracted state of the country. Would it not be sufficient if you guarantee that I am trying to get into useful country, without saying exactly where or by what route?

Apart from collecting for Bees, who remained his primary sponsor, the other objective of the new expedition was to explore the more remote parts of the Brahmaputra, the river whose name had excited his imagination when he first heard it mentioned when he was a child. And however great the problems of entering China may have appeared in 1912, by 1913 he was *en route* to the wilderness again.

Before leaving for China, K-W had been through a crash course at the Royal Geographical Society in surveying and map making. This entailed trying to absorb a highly complex and technical subject in a few months, when professional surveyors and cartographers learned their craft over years. Ronald Kaulback, who travelled with K-W in 1933, said that he had 'to work like a slave with maps and calculations, theodolites and plane-tables' for about two and a half months when he took the RGS course.

A theodolite alone is a highly complicated piece of equipment, especially in the hands of a beginner, and it was vital to be able to take it apart, reassemble it, even repair it if necessary. He needed to carry out theodolite traverses. In a military manual on the principles of map making which would then have applied, the uses of theodolite traverses were listed as taking the place of triangulation in flat or forested areas where triangulation itself is impossible; supplementing existing triangulation in areas where the more normal method of 'breaking down' is impossible or inconvenient; and providing a 'control' for subsequent mapping.

In fact it became increasingly difficult for K-W to use a theodolite without arousing suspicion. As he explained to Scott Keltie, 'Owing to the military activity out here I am regarded as an arch-spy and have to be careful.' In October he again wrote to Scott Keltie, this time from Atuntsi:

> I have been able to do practically nothing in the mapping line here for good reasons; in the first place my botanical work has demanded all my attention, and in the second place there has been a rather serious row about my theodolite, which at one time nearly involved my removal from the province.
>
> I only use the theodolite occasionally and very secretly at night to try and get a good *average* rate for my watch which I can use on the trip.

His plan was to attempt to cross Eastern Tibet, and if he was to have any chance of succeeding he had to avoid incurring the wrath of the local Chinese officials by spying out the terrain with his theodolite. However, he was able to make a series of plane-table sketches.

The plane-table was a rather more discreet piece of equipment — a drawing board mounted on a tripod which could be rotated in a horizontal plane. It could have been mistaken for some kind of artist's easel, although the additional pieces of equipment which went with the relatively plain and simple drawing board ran the risk of arousing suspicion — the alidade which held the sight rule, the clinometer, the instrument used for measuring slopes and elevations, a scale, compass and field glasses, the fine needles for marking fixed positions. About the only things that could have passed him off as an artist were the pencils, penknife and india rubber. However, the plane-table did enable him to get down the features of the terrain he was surveying so that they could be worked up by the professional map makers at the Royal Geographical Society.

Less obtrusive were the aneroid and boiling point barometers used to calculate the height of mountains. Neither was absolutely accurate, but certainly good enough to give a very fair notion of heights in uncharted country. An aneroid barometer carries a scale of heights from sea-level. The mercury responds to atmospheric pressure, and from the calibration it reaches it is possible to calculate height.

K-W more usually used a boiling point thermometer. Atmospheric pressure at different altitudes determines the temperature at which water boils. If you take that temperature height can then be calculated against tables. He also carried a keyless half-chronometer, a beautifully made watch that would withstand rough treatment and continue to keep accurate Greenwich mean time. From this he was able to calculate longitude from the angle between the meridian at the point where he was, and the meridian set at Greenwich Observatory.

The whole business of surveying and map making did not end with taking and noting instrument measurements, it also involved considerable calculations. It was time-consuming and had to be set aside when the demands of botany and plant and seed collecting became too demanding. In later expeditions he was able to turn the map making and surveying over to the men who accompanied him, like Kaulback, Cawdor and Clutterbuck. After the Second World War he dropped that side of his work completely. But in 1913 he was bubbling over with energy and ambition and despite a becoming modesty was determined to reach the top of his profession, and he busied himself with the instruments loaned to him by the RGS.

He wrote to Scott Keltie that he had been making compass traverses from Atuntsi to his camps on the Mekong–Salween and Mekong–Yangtse divides 'with latitudes of the camps and azimuths of two of the snow peaks'. The azimuth observations, which involve measuring altitude against a star and the horizon with the aid of a theodolite, meant that he could use the instrument at night. He hoped his work would enable him to 'draw a very rough map of the district (about a degree each way on the two miles to an inch scale), but I don't think it will be of the least value ... As a matter of fact I have amused myself in this way chiefly for practice with the prismatic compass, estimating distances, calculating altitudes, and using the theodolite for star work and rounds of angles, as I hope to do a little useful work on my journey.'

The journey during which he hoped to do this useful work for the RGS had originally been planned to take him to the fabled Brahmaputra Falls, but he abandoned this idea when he heard that the eminent explorer Lt-Col F. M. Bailey was even then on his way to the area, '... and he will have got to the falls by this time – he is not a man to fail,' he wrote.

In his second letter to Scott Keltie, K-W admitted that his first scheme was 'too big an undertaking for me ... I don't think I shall attempt it, though I should like to see the Tsanpo [Tsangpo] above the falls.' He went on to outline his new plan:

> I have heard something about the province of Pomed, which I am particularly aiming for – of warm valleys, big lakes, big rivers and high mountains, and to get there we must cross Tsarung – a cold journey.
>
> If we reach Pomed in good time, I shall try and push up the Brahmaputra, and if all goes well attempt to reach Lhasa by this southern road. But as I may fail in the first fortnight, it seems best not to talk too much about it.

As it turned out, the attempt to get into Tibet was little short of a disaster. Any hopes he had of the Tibetan officials being more amenable than the Chinese were soon dashed. On the last day of January 1914, again based in Atuntsi, he wrote to Scott-Keltie of returning, 'after 92 days of effort and complete failure in my journey. However, we did our best, better luck next time; which if I survive a serious rowing and am allowed to stay in Yunnan, will I hope be next year. What I thought was an excellent opportunity to cross Tibet proved, without assistance from Lhasa or India, the worst possible time on account of the Chinese trouble.'

The Tibetan government refused to allow him to travel in their country, and an attempt to cross the Irrawaddy was frustrated by a Chinese official who banned the local tribesmen from carrying his baggage. He tried to escape by staging a three-day dash down the Salween, but the most appalling weather closed in and the few ill-clad porters he had with him refused to continue. He had to retrace his steps and was tracked down by the Chinese who once again forbade the porters from carrying his loads. Without using actual force, the Chinese were driving him back to Yunnan where they could keep a close watch on him.

The Making of a Plant Hunter

After that everything miscarried. We crawled on down that awful Salween valley hoping to get across lower down, but it rained and snowed day after day, all the passes on both sides were hopelessly blocked, and the outlook was most dismal. I got fever, the Lisus were very wild, my Lisu interpreter bolted in the night, we had hardly anything to eat (the country only produces a little maize and buckwheat, with here and there a small amount of rice), and I hardly knew what to do. However at last we got down as far as Latsa, and then the only thing to do was to come back. When we reached Chamutong on December 22, the Chinese practically made me a prisoner on parole; but as it snowed again, blocking all the passes to A-tun-tzu, I was eventually allowed to go north into Tsarung again. I hoped by this time the trouble would be over and the Tibetans allow me to pass, but it was on the contrary worse than ever, the Chinese, so it was said, having gained a foothold in Tsarung a few days north; consequently we were hustled out of Peitu in double quick time, before we had time to breathe, and reached A-tun-tzu on January 29, very fit after our saunter through glorious Tsarung in splendid weather – a great contrast to the miserable time we had for a month amongst the Lisus.

By now news had reached him that Bailey had reported there was no great waterfall on the Brahmaputra, or even a series of waterfalls as suggested in local legend. However, K-W still did not discount the possibility of such a natural wonder on at least one of the great rivers that cut their way through the Himalayas. He believed it would be found in Pomed, which he had not yet reached. 'But of this I am sure, Pomed is a country where we shall find some wonderful things and solve some of the Indo-China puzzles, and I will go there if I die for it, some day,' he declared.

His letters to Scott-Keltie tended to be extremely self-deprecatory, and revealed a sense of guilt in not, in his opinion, making the best use of the equipment and expertise provided by the Society. 'I am awfully sorry I have done so little after all the assistance the RGS has so kindly given me ... I blundered fearfully I know, but I understand better how to succeed at the next attempt ...' But the expedition was far from being a wasted effort. He introduced five first-class new rhododendrons to Britain, including the incomparable yellow-bloomed *Rhododendron wardii*, which has subsequently sired some superb hybrids (one of the most recent — 'Tidbit' — probably would not have pleased him. It is a cross between his beloved *wardii*, and *Rhododendron dichroathum*, a plant he particularly disliked). The other four species were *Rhododendron charianthum*, *R. melianthum*, *R. aganniphum* and *R. campylogynum*.

He also forged a closer relationship with the remote, wild hill people, who in turn developed a deep respect for him. As for the Chinese, although he was unpopular with the officials, K-W got on well with the ordinary Chinese, including landowners and businessmen, and on more than one occasion was handsomely entertained. At one feast he was required to consume prodigious quantities of food and *shao-chiu*, a fiery drink which he found tasted like methylated spirits. When the meal finally ended he fell into bed and 'stayed there quite some time'.

The gramophone which he had taken along for his own entertainment soon became an essential piece of equipment in his dealings with the local people, who found it an endless source of delight and astonishment. It proved a particularly good aid in photographing the normally camera-shy locals. When he was passing through a small town on his way to the Salween he set up the machine in the middle of the street. A crowd soon gathered, and he arranged them round the gramophone in a suitable group for a photograph. 'Then I turned on the band, and dumb applause following, no sooner had I got them thoroughly wrought up than I whipped out my camera and took several photographs of the enchanted crowd. When the banjo tinkled the children laughed with glee. How they did enjoy that wonderful box to be sure, and I had to play every tune before they were satisfied.'

On another occasion he arrived at a small monastery on the Mekong where he met a lama from another house who had heard about 'my wonderful singing-box, and nothing would satisfy him but that I should at once bring it out. Though it was late in the evening I complied, and all my porters, girls as well as men, in spite of their heavy day's march, sat round in a ring to listen; while from ten o'clock to midnight I held the company entranced.'

Because of the trouble he had caused the Chinese authorities by what they considered to be his illegal journey in Eastern Tibet and on the North-East Frontier, he was virtually banished from China, and had to base himself on the Burma–Yunnan border at Hpimaw Fort. It was a frontier fort in every sense: a small building constructed from stone and wood, its walls reinforced with brushwood and turfs with which to absorb rifle and machine-gun bullets. There was a central courtyard that could act as a shelter for refugees and animals if the building came under attack. He described its setting, which was of a wildness and grandeur that always appealed to him, in his book *In Farthest Burma*:

> Gray granite, knotted and corrugated, pleated and crumpled into bewildering tangles, and again hacked through and through by destructive storm waters; stark cliffs of limestone overshadowing the valleys; slopes here clad with rain-drenched forest, elsewhere so steep and rocky that nothing but rank grass and desperate grapple-rooted trees find foothold in the short soil; and on a bleak, windy shoulder where a spur, sweeping down from the crest of the range, has broken its back and tumbled in agony to the deep valley of the brawling Ngawchang Hka, blocking the path to China, stands Hpimaw Fort.
>
> From the commandant's bungalow just below the fort itself you look across the marble-clouded valley, where invisible villages are snugly tucked away in the folds, to the grey-blue mountain ranges of the 'Nmai Hka, crowned by the gaunt mass of Imaw Bum, white-furrowed where the snow-choked couloirs spread fingerwise into the valley. Behind the bungalow the darkly forested slopes of the main range rise abruptly. The path to China follows the spur from the fort, climbing sometimes steeply, sometimes gently, now perched on the crest, now slipping over and traversing one or other flank.

Here, in the company of one other European, the commanding officer, and one hundred Gurkhas, he botanised in an area which was to become very much his own in the years to follow. It was a country rich with flowers — alliums, nomocharis, enormous rhubarb, primulas, and poppies like *Meconopsis wallichii*, which grew seven feet high and sported great spikes of pale purple petals set off by a massive boss of golden anthers.

Having been forced to abandon any further exploring and collecting in China and Tibet, in 1914 he shifted his attention back to the mountain range of the North-East Frontier, and a reconnaissance of the Burma–China frontier.

His travels took him into the Maru country. This tribe occupied part of the valley of the upper Irrawaddy, and had developed from an intermingling of the Burman and Tibetan races, but had a language closer to Burmese than Tibetan. Since independence the Maru and the other tribes in the Northern Territory of Burma answer to being Kachin.

In K-W's time they lived a fairly primitive life, but he developed a great fondness for the region and its people, especially the girl porters, who, he said, 'were always merry and bright, kept the men in a good temper and were not ill-favoured to look upon'. K-W was always susceptible to a pretty face. When he was last on home leave in Cambridge he had fallen hopelessly in love with a sixteen-year-old girl whom his sister, Winifred, described as 'lovely as a fairy-tale princess, with a cascade of moon-fair hair hanging about her shoulders'. He proposed to her, which was something he was always inclined to do the moment he fell in love. She refused him.

Although not a prude, K-W was astonished by the nonchalant promiscuity of the unmarried Maru girls and boys. In one hut where he was staying his bed was set up close to the hearth where the girls slept, and were visited during the night by the boys. 'Courting carried to its logical conclusion', was how he described it.

Despite some severe attacks of fever, and the death of his little terrier, which he had named Maru, the 1914 expedition was a happy one. Being quite cut off from outside communication he was unaware that the First World War had broken out, and he remained in ignorance until he reached Fort Kawnglu, another remote outpost, 'with six weeks' growth of beard, dirty and haggard, my clothes worn out, my boots flapping, my hair long'. Like other young men at that time his sole ambition now was to play his part in the war. With as little delay as possible he set off for Fort Hertz at Putao, an important military establishment built in 1911 when Britain annexed Burma. From there he intended to march to Myitkyina and join up. Already tired from his travels, the rush to Putao proved too much. Leeches and flies took their toll, and by the time he reached the fort he was in the grip of fever, which laid him low for six weeks.

When he recovered, K-W was able to join the Indian Army and was posted to Victoria Point at Tenasserim in Lower Burma. Writing to Mr Hinks of the RGS, he said:

> This is a weird little place, which has attained a certain local notoriety on account of the war. There are only two white men besides myself, but we receive visits from the patrol boats which are searching the archipelago for gun-runners, from one or two rubber planters up country, and from the tin mining people on the Siam side across the estuary.

Arthur Robert Hinks, who became one of K-W's chief correspondents, was Assistant Secretary of the RGS and editor of the *Geographical Magazine* until 1915, when he became Secretary of the Society. In another letter posted on 2 December 1915, K-W complained:

> 'Except for a certain amount of work soldiering with the Goorkhas [sic] garrison here — as much as I can get them to do as they are not under my orders, and moderately lazy — I have hardly any war work, and have to spend my time in this way. It is rather sickening, as I had hoped to be posted to a regiment. Our garrison is about 60 strong, half being the wireless guard and half on outpost duty, each lot under a native commissioned officer. I have them out for parades and skirmishing when I can get them!

But the letter did end on a slightly up-beat note, perhaps because another compatriot had joined them: 'We are four white men here and don't have a bad time.'

Nevertheless he was frustrated. With his regular plant-collecting fee from Bees, the small amount he earned from writing, and a good deal of prudence, he had been able to save some money, only about £300, which he planned to invest in a major expedition to Tibet in 1915. In a letter home in 1914 he said the trip was fixed up, but that may well have been wishful thinking.

> Probably two other men are coming with me [he wrote]. We shall have our own caravan and try to cross about half Tibet and come down to India over the Himalayas. The objects of the journey for me are to explore and map all that unknown part of S.E. Tibet towards and south of the headwaters of the Salween, and to seek a strange tribe that the Tibetans say live in trees in that country! To collect plants and continue my observations on the distribution of the Indo-Chinese plants, particularly Primulas, with a view to discovering the connection between the Himalayan & Chinese floras; to photograph the tribes, scenery, etc., and collect rock specimens to help in tracing the trend of the mountain ranges from E. to W. in Tibet, to N. & S. on the China frontier; and generally to further our knowledge of this interesting country. We expect to take the best part of a year, if successful.

K-W had gained the confidence to mount what would have been a major expedition of scientific research and exploration. Looking back on his achievements in 1913 and 1914 he had reason to be self-assured.

> ... I collected about 1,000 species of plants, many of which I am confident are new species; some of these will now be introduced in England, as hardy plants, for the first time. In the winter I mapped my route across a province of S.E. Tibet & down the upper Salween, making observations on the Tibetans, and the Lutzw & Lisw [Lusu and Lisu] tribes,

some of which are I daresay original, and secured a good series of photographs of these people. Geographically I discovered a new snow mountain & fixed its position roughly. Our camps on the snow range near Atuntsi enabled me to make some interesting observations on the unusual glaciers of these mountains, and these observations together with observations on the climate, geology and distribution of plants have led me to formulate a theory to account for the connection between the Himalayan & Chinese floras, on which I am busy now. Finally, various climatic and other observations at high altitudes have enabled me to collect some notes on scree and alpine plants, together with the theories of distribution, control factors in distribution, and so finally to this hypothesis of Indo-China distribution in general. If I can make the journey we are going to attempt I hope to be able to test this theory in many points, and gain a much more comprehensive view of the country, arrangement of the mountain chains, and limits of the two great climatic areas which will afford the clue to the whole problem.

One part of his theory was that the long history of the formation of the Himalaya could be traced through the distribution of plants; how the numbers and characters of the species had changed and adapted as their terrain and climate had altered during the great convulsions that had squeezed the mountain ranges out of the face of the earth. The distribution of plants, particularly those like some lilies and other bulbous plants that were used for food, could also give a clue to the movements of people.

It is not difficult to imagine K-W's frustration at having to abandon his plans. With success he was sure that the expedition would have guaranteed him all the financial backing he would have needed in the future. As it was, he was isolated in a tiny Burmese garrison with little to do but play at soldiering with rather reluctant native troops, and carry out censorship duties, while thousands of miles away the real war raged without him.

It was not until December 1916 that he got a break. He was

posted to Mesopotamia as a lieutenant in the 116th Mahrattas with the Indian Expeditionary Force 'D'.

He wrote to Mr Hinks in February 1917: 'I have been here two months with my regiment, and in spite of the monotony of the desert — we are camped on the edge of it — I find such parts as are inhabited very interesting indeed. There is an almost uncanny resemblance between the Arab architecture here, and Eastern Tibetan architecture as seen in the Mekong and Salween valleys.'

There was relatively little action, however, and at the end of the week's work he was able to go for long Sunday rides in the desert to Zabair and Shaiba. He found plants to study and birds to spot in the palm groves, and when darkness came he would lie in his tent listening to the jackals 'making night hideous', and wistfully think of his beloved Himalaya. 'Of course one misses the mountains — there seems to be a vast gap between earth and sky.' Tibet, and its great protective rampart of mountain ranges, was scarcely ever out of his mind. If the war was over by then, he planned to start his great Tibetan journey in April 1918.

In November 1917 he was in Baghdad, mainly involved in routine army work, which he found tedious in the extreme, and although he left the army in 1918 with the rank of captain, the war for him had been thoroughly unsatisfactory. He felt he had been under-used. He was, after all, a man of action, a man of decision, who had learnt the art of leadership in the most rugged, lonely and difficult circumstances possible. It seemed to him that the professional soldiers had quite failed to recognise his qualities.

His letters home give some idea of his military life. Writing from Baghdad in October 1917, he describes dining with 'my Chaldean friends' and trying 'to teach the girls to dance. Camille is not an ungraceful girl, but Antoinette is like a little hippopotamus.' The nearest thing to action he records in these accounts of his routine life was when he was put in command of a company of two hundred men 'who were told off to line a portion of the route along which 1,000 prisoners from Ramadie with six of the field guns we took, were to march through the city — as a lesson to the Arabs. I pranced along on my charger, no doubt looking no end of a warrior, but feeling quite the chocolate soldier.' He had had no part in the battle that produced the prisoners and guns.

In the middle of November he wrote home, again from Baghdad, remarking, somewhat bitterly: 'You are seeing much more of the war than I am, with aeroplanes dropping bombs on your heads every day. I have been nearly a year in Mesopotamia and I haven't yet seen a shot fired in anger. Of course I am lucky not to have done so, but it makes one feel a fraud nevertheless, and I should like to strike a blow at the hated Huns.' At one point K-W believed he would be sent to the Dardanelles where the Allied troops and the Turks were fighting it out in hand-to-hand combat. But it did not happen and after four years in uniform he left the army unscathed and unblooded.

There is no doubt he was disappointed by his military experience, and that he felt aggrieved at being undervalued. It was a feeling he was to experience again in the Second World War, and from time to time during his travels when he went for months on end without any communication from friends or sponsors. It was a sense of rejection that was inevitably aggravated by the loneliness of his chosen life.

4

Back to the Flower Trail

IN 1918, at the age of thirty-three, K-W was in Simla in India desperate to get to China and mount the Yunnan–Tibet expedition he had planned for so long. But every effort was to be frustrated by the turmoil in China, and a wave of anti-Western feeling that swept through the country with the belief that the Allies had sold China out to the Japanese.

Japan had declared war on Germany for the sole purpose of acquiring Germany's Chinese territory in the province of Shantung. Not content with that, in 1915 Japan had presented Yuan Shih-k'ai with the notorious 'Twenty-one Demands', which insisted upon China recognising Japan's sovereignty in Shantung, Manchuria, Inner Mongolia, and the coastal area which faced Taiwan.

In 1917, after Yuan's death, China declared war on Germany, hoping to recover the former German territory, but in 1919 the Treaty of Versailles divided the German colonies between Japan, France and the British Empire. The decision sparked off the May Fourth Movement and serious rioting in Peking and Shanghai.

Under these circumstances a Western expedition in China and Tibet was out of the question. Once again K-W had to abandon the enterprise and content himself instead with returning to Hpimaw and Imaw Bum to 'botanise quietly'. Although it was on a modest scale, the 1919 collecting trip produced a fine crop of twenty-six rhododendron species, which included three of his special favourites: *Rhododendron myrtilloides*, a dwarf which he valued for its tiny, neat, myrtle-shaped leaves; and two tender epiphytic species, which he described lovingly in *Plant Hunting in the Wilds*: 'High up on a moss-bound fir hangs a milk-white cloud of *R. bullatum*, whence from time to time there flutters down a fragrant corolla, like a broken butterfly; and clasped in the fork of another tree is a small bush studded with button flowers which gleam pale golden amongst the silver-plated leaves – *R. megeratum*.'

In 1920, after an absence of seven years, he returned to England. His mother was no longer living in Cambridge. She and Winifred had taken rooms in a small hotel in Bayswater. Britain was recovering from the horrors of the war. The flapper era brought an almost hysterical gaiety to London, and K-W fell in with the mood enthusiastically. He was an attractive man. At only five feet eight inches he was quite short, but lithe and wiry. He had arresting sea-green eyes, piercing at one moment, dreamy the next. Even at quite an early age his hair, which had darkened with adulthood, was turning to the pure white that so distinguished him in later life. His voice was modulated and beautifully clear.

K-W enjoyed the company of women and fell in love easily, often signing his letters to his women friends 'your little friend', and not infrequently proposing marriage. On the homeward-bound boat he had fallen in love with a girl called Alice, and while, for once, he had not proposed, he was determined to woo her, but her family was horrified at the prospect of her becoming involved with an impoverished plant collector. His love letters were intercepted and the relationship ended. On the rebound he proposed to a girl he met at his mother's hotel. She refused him.

His enjoyment of London social life was marred by his concern for the straitened circumstances of his mother and sister. He felt guilt-ridden by his inability to provide for them adequately or at all. He could have returned to Cambridge and an academic life, or worked in the herbariums at the Kew or Edinburgh Botanic Gardens, or at the Natural History Museum's Department of Botany. He did consider these options carefully but knew perfectly well that he would never be able to resist the attraction of the mountains and his Asian wilderness. However, an apparently perfect solution did present itself: he would start his own nursery garden to grow and market the plants he collected and introduced into Britain. It would be unique in as much as no other plant collector had had his own nursery, but normally collected for companies like Bees or Veitch or for institutions. He would be able to keep it stocked with marvellous new plants from future expeditions that would be financed by the business.

With part of his small savings he bought a partnership in a business in Devon which amounted to little more than a market garden. At first all seemed to go well, and he enjoyed the summer months of 1920 learning how to grow tomatoes, picking beans, and taking the produce to market in 'a ripping little Ford van'. But K-W was never a businessman, and trustingly left the development of the nursery garden to his partner, who was a heavy drinker and inept manager. Bills were unpaid, and the accounts fell into chaos. In the end the ruins of the partnership had to be put into the hands of a solicitor. His advice was that all that could be done was to wind up the business. Thus ended K-W's one and only attempt to enter the commercial world.

In 1921 he finally managed to get back to Yunnan in China. Writing to Mr Hinks at the RGS from Yungpeh towards the end of December, he reported: 'I have had quite a successful year, spending the summer and autumn at Muli on the Litang River, and working on the range immediately to the west, between the Litang and the Sholo Rivers.'

He was again collecting for A. K. Bulley, the founder of Bees, and working on a relatively low ridge, averaging only about fifteen thousand feet, and out of the range of the snow fields of Tibet. Because it was so dry there was little timber, but the flora was magnificent, yielding twenty-five rhododendrons, some forty primulas, and four blue meconopsis, one of which, he said: 'is I think the finest flower I have ever seen'. He was, he admitted, less successful with his mapping work. In a letter to Hinks written on 1 December 1921, he explained:

> To begin with the weather was very bad – heavy rain and thick cloud all summer, and snow in the autumn. There

was no possibility of measuring a base, so I used an astronomical one, and fixed three peaks from which I intersected the snow peaks and got their altitudes by vertical angles. In the autumn I was in camp 5 weeks and we had 4 fine days! Consequently the plane-tabling became rather sketchy. However I got in a little of the glaciated range, and the Muli valley, and fixed the snow peaks — which I think are new. As soon as I have finished this map I will send it to you with a paper I have written on the glaciation, which may interest those who are keen on glaciology. But of course the map is not good, it is merely better than nothing. I suppose one can hardly expect to make a really good map at the first attempt. Anyway it was tremendously good practice and good fun too, except taking angles on the tops of 15,000-foot peaks in November, which was a cold job; and next year I hope to do better.

The journey was made additionally interesting by the fact that he crossed Yunnan by remote, little-known roads from north to south. Starting from Lashio he went by the Kunlong ferry, Shunning and Menghwa to Tali. From there he made his way to Yungpei, and due north to Yungning and Muli. On the return journey he measured a base on the little plain of Yungning, plane-tabled the immediate neighbourhood, and carried a traverse down to Yungpei, as this road was not shown on the contemporary map of Yunnan. It proved to be an important route.

In 1922 he undertook an even more arduous journey from Likiang to Hkamti Long, collecting in the Tibetan marches, eventually returning to Likiang by way of Tengyueh, Bhamo, Tali, Yungning and Muli. This was country that he adored, and in his book *From China to Hkamti Long* he wrote excitedly: 'To traverse the furrowed crust-belt, from the Yangtse in the east to the Irrawaddy in the west; to cross the great rivers and climb the great divides, was my ambition ...' Unfortunately fever intervened and brought him dangerously close to death.

Back to the Flower Trail

Again in his account *From China to Hkamti Long* he describes the awful experience. Illness struck when he was camped on a sandbank:

> At dusk a shower came on, and I set up my little sleeping tent, and crawled in, feeling too unwell to take any supper. All the following day I lay there, hoping that the next day the fever would leave me, and I could go on. The sun was terribly hot, so I made the Nungs pluck banana leaves with which to cover my tent. It was impossible to take any food, but I drank copious draughts of cold water from the river, to quench the fever.
>
> By evening I could hardly stand. Four marches would have taken me to the Krong-jong Pass, from where we could look down into Assam, and it was sad to have to return before we got there.
>
> However, it was the only thing to do, and that evening I gave the word to return. The men were to make a bamboo stretcher and carry me back to Fort Hertz as quickly as possible.
>
> But here a difficulty presented itself. If the porters carried me, obviously they could not carry the loads as well. The Mishmis were all down by the river, fishing. I told the men, therefore, to jettison half the baggage, and send the Mishmis back for it, and, after another weary day of raging thirst, we started south.
>
> Everything went wrong from the start. The men only marched about three miles, and then the stretcher was dumped down by the river under a burning sun. For hours I lay there – the porters had gone back for the luggage. In the evening they returned, and carried me to a temporary shelter they had built, and lit the fires; but I woke in the night, crying for water. I had had no food for two days. The men smoked opium incessantly – before breakfast in the morning, and after supper at night. I watched them with curious fascination. On the following morning my guide told me, in his broken Burmese, that the men could

not carry me over the boulders by the river – it was too difficult for them.

But there was a path through the jungle, over a spur, and so we went that way, and in the afternoon reached the ferry where the Mishmis were. We crossed on the flimsy raft, and at dusk reached our previous camping-ground.

The night dragged on. From time to time a man rose up to put more wood on the fire. I was feeling weak, for it was now three days since I had eaten any food.

Dawn at last, chill and misty. The Nungs were in no hurry, and I could only whisper instructions to my guide. But at last we started, and leaving the river and jungle behind, came out on the plain itself. The sun beat down fiercely as we came, early in the afternoon, to the village of Namsai. Here I was able to eat two raw eggs.

On December 11 we marched across the plain to Manse Kun, and came to the bank of the Nam Palak, where we camped. It is a common fallacy that sand is soft. True, you can mould it to the required shape, but in point of fact it is cruelly hard to sleep on. Once more I passed a nightmare vigil.

How thankful I was to see the grey dawn on December 12! How terribly slow the men were over breakfast, which did not interest me. A dugout was drawn up on the beach, and presently I was laid in this, and off we went.

Arriving at Fort Hertz, I was taken straight to the Civil Hospital, where the doctor and his assistant kindly attended on me and made me comfortable. On the following morning Captain Cousins of the Military Police, who had taken Leeming's place, came across and insisted on having me moved into his bungalow, where I was well looked after by Cousins and the Civil Surgeon.

With this failure ended my second attempt to march overland from China to India direct; but I had at least come much nearer to complete success than on my first venture in 1913.

This was not his first bout of fever, but it was certainly one of the most serious. By now fever was in his blood, and he was never to be free from it. At only thirty-seven, physical hardship and recurring bouts of fever had taken their toll. His whitening hair and lined face made him appear older and more timeworn than most men of his age.

As well as the disappointment of having to abort his journey, K-W also had to face the grief of his mother's death in June 1922. When he had said goodbye to her he had no idea that she was ailing, but shortly after he left she fell ill and died following a long illness. It took six months for the telegram from Winifred bearing the news to reach him.

However, Winifred remained in the small Bayswater hotel and K-W still considered this home. He remained as devoted to Winifred as ever, but at this time there was also someone else in his life — a strikingly beautiful young woman of twenty-three, Florinda Norman-Thompson, who was to become his wife. They had met during his last leave, and he had fallen in love. His proposal of marriage had been turned down, but on New Year's Day, 1922, she sat down and wrote a letter asking him to marry her. It was typical of her. She was a strong-willed woman who liked to make the decisions. In a long, rambling, and sometimes confused, letter she wrote:

> My Dearest,
> I wrote you by last mail, and have since had your letter of December 9th. This shows you have not had any letters from me for a long time, you seem to get very few of them. I hope some have turned up since then. As this is rather an important one (to me) I am sending it out to Kenneth [Ward] and asking him if he can redirect it any better than Cooks can, or can find, or encourage them to find, any more likely means of tracing you.
> In case you did not get last weeks letter I am making this an amplification of it.

My Dear, I wrote to ask if you would marry me when you came home. I'm afraid I made it a rather short note, but there is so much that I just have to hope that you will understand. I thought last year when you went away that you were doing something extraordinarily brave, but I did not know quite how brave it was.

I know you are anxious about me for more than one reason, but you had so sure a faith that you would be unafraid. I think in a way that helped me then. I know it helps now. Dear, I could not do anything else. [She was referring to her refusal to his proposal of marriage.] I could never leave a chance to marry you for anything but the proper reason. If I had just married you out of cowardice, as I wanted help and you offered it, or out of sympathy, as I thought I could make you happy, nothing would have been right. So please be as wise now as I was, and if you do not any longer want me for your wife say so. It won't be any worse for me than it was for you, and we will always be friends.

But if you still want to marry me will you let me know as quickly as you can. I do not want, I most particularly do not want you to alter any of your plans, or to hurry your work in the very slightest degree. I feel quite sure I can wait for you just as well as you were ready to wait for me; and I feel quite sure that I can belong to you properly now if you want me, and that we two can help each other to make happiness in the world, and to see things in their proper proportion.

I had to wait 'till I was sure, Dear. You see being married with any kind of mental reservation would be unfaithfulness really – wouldn't it.

You understand why I went on as I did last year, don't you? If you don't I can't explain any better than by saying I could never put a 'job of work' down while I thought it could do any good by keeping on.

If we are not going to be married, don't forget that we are always friends.

Please don't worry about me. I am well and have work to do.

Only, of course I shall want to hear from you as soon as possible. I planned this letter to you this morning when I was out walking on the Thicket. I don't care what the calendar says, it was really the first day of spring. There was a heavy mist in the morning, and it broke before the sun as I was walking, and I saw that the fields were full of green things beginning life, and that the outlines of the buds on the trees were softening. I hope that your lovely Chinese winter has been kind to you to date.

What ever happens, Dear, you have given me very much strength and faith and friendship.

This is not really the letter I planned for you – but you will perhaps understand.

For a love letter she ended oddly: 'A fine night, citizens, and all's well. Florinda.'

Her family had its roots in Ireland, and she had inherited the Celtic character in full measure – that volatile mixture of incorrigible romanticism and impossible ambition. She impressed people in sharply differing ways. To some she was a hard, uncompromising 'bluestocking', apparently devoid of emotion, who would astonish them by collapsing in apparently inconsolable grief whenever she saw K-W off on his expeditions. To others she was a woman of irresistible charm and persuasiveness, who could make even her most eccentric plans seem plausible. That she was strikingly good-looking there was no doubt. She was quite a bit taller than K-W, had fine-boned aquiline features and a mass of blonde hair, which was always put up. She dressed in an idiosyncratic way, but with tremendous flair, in long dresses and flamboyant picture hats, while the fashion of the time was for short skirts and cloche hats. One person who knew her well said: 'She would always be noticeable in a crowd, not in a way that people would laugh at her, but they would say "that is a remarkable woman".'

K-W named a number of plants in Florinda's honour, probably

the best known being *Primula florindae*. Its cascades of clear yellow bells must have reminded him of her hair. Perhaps that was also true of the pale yellow blooms of the rare woodland poppy, *Meconopsis florindae*.

They were married in Kensington Register Office on 11 April 1923, although K-W was still suffering seriously from the after-effects of fever, and Florinda's doctor had warned her that she would have to nurse him as an invalid.

Florinda was not a rich woman, and, as we have seen, K-W's finances were always of the most precarious, but she insisted on living in style. When, after they were married, they moved from London to Hatton Gore, a Georgian house in Harlington (since destroyed to make way for part of Heathrow Airport), they had a butler and cook, and several gardeners. She once astonished some dinner guests by explaining that the silver plates they were eating off were more economical than china because they did not break.

Florinda now directed her considerable ambition and energy to advancing K-W's career. She swept him off to Ireland where he met Lady Londonderry, an enthusiastic gardener and patron of plant hunters, who was to buy seed shares in his future expeditions. It was Florinda who introduced him to Jonathan Cape, who published his books for many years, and at Hatton Gore she organised a busy round of dinner and weekend parties, inviting people who might commission him to lecture or write articles.

To those who knew Florinda and K-W at that time they seemed a perfect match – the now established explorer, and the beautiful, talented wife devoting herself to promoting him. Only the most perceptive could have recognised that in reality it was a sad mismatch.

While Florinda was busy championing K-W, his mind was on very different things: he was absorbed in planning and organising what was arguably to be the most outstanding expedition of his career – the long trek to solve the riddle of the mysterious Tsangpo gorges.

5

To the Land of the Great Gorge

EARLY in 1924, K-W set off on this expedition, which was to prove of great geographical as well as botanical importance. The objective was the gorges of the Tsangpo, the Tibetan river which becomes the Brahmaputra once it has burst out of the rampart mountains of Tibet. Just before the First World War, Bailey had explored the upper reaches of the Brahmaputra, hoping to find major waterfalls. They did not exist. He did not make much progress into the Tsangpo gorges, but nevertheless was inclined to dismiss as mere myth the belief that there were a series of huge falls in the gorges themselves. The existence of falls was certainly enshrined in Tibetan lore, and they were said to be guarded by individual spirits. There was also the fact that at Lhasa the Tsangpo flowed at twelve thousand feet, but when it emerged from the Abor Hills it was only one thousand feet above sea-level. It seemed logical that such a drop must include at least one mighty waterfall. The answer was in the fifty unexplored miles of the gorges, and K-W intended to settle the question.

The other and equally important purpose of the journey was to 'collect plants in a region which was an even greater mystery from the botanist's point of view than from the geographer's. We intended to make a collection of garden plants, and so introduce them into Britain, which is the world's temperate garden.'

Kingdon-Ward was accompanied by Lord Cawdor, who put up part of the finance for the expedition. The remainder came from

Frank Kingdon-Ward

Bees and other seed share investors, and the usual help in kind from the RGS, and from K-W's own slender resources. The explorers sailed from England to Calcutta, travelled by train to Darjeeling, and then overland to Phari, the first town high on the plateau of Tibet.

The two men could scarcely have been more dissimilar. In the first place K-W was fifteen years the senior, and while he did have a good sense of fun he was first and foremost a serious explorer and plant collector, and always dedicated to the job in hand.

Lord Cawdor also had a genuine interest in exploration, and devoted much of his time during the expedition to surveying and map making, relieving K-W of that task. He was also a reasonable photographer, and hoped to make himself useful by taking on responsibility for that. But at twenty-four he also sought adventure; there was a touch of the gung-ho about him. He loved to stride out over the wild mountains, or to ride a rough little pony over the Asian moorland. His journals are rather endearingly peppered with the language of the English public school and the naval officers' ward room; he had served in the Royal Navy during the First World War. His vocabulary on some occasions would certainly have been disapproved of by the Race Relations Board. His journal entry for 16 March 1924, for instance, observes: 'The local coons made a beastly wailing noise and beat drums far into the night celebrating a Hindu Festival (spring fruitfulness etc. I fancy). Also the jackals howled all night.' K-W's view was somewhat different. On 17 March he recorded: 'We met Tibetans on the road returning to Phari – it was good to smell the sour, rancid smell of a Tibetan again.'

Before starting in earnest on the expedition, K-W and Cawdor spent a short time at Gyantze, which had a tiny British civil and military community. At first K-W was unwell, suffering from an ulcerated tongue, but Cawdor entered with gusto into the sporting life of the post, as his journal reveals:

> 6.4.24: Sunday, Gyantze: Football at 0900. I played for the Tibetan team against Sikhs. W [K-W] refused. We went to

the fort at 16.30 and had tea — afterwards we played lawn tennis on quite a good hard court made of local cement.

Hislop and I versus Cobbet and W. Scotland won after a struggle — at least as far as I remember — but I may be wrong.

K-W (Cawdor refers to him just as W throughout his journal) even recovered enough to play polo, for the first time in his life.

His diary entry for 10 April records: 'Gorgeous morning. Played four chukkas of polo at nine on bobbery ponies. Great fun. First pony wouldn't gallop, second wouldn't go near the ball.'

On 11 April they got under way, their baggage carried by a collection of ponies, mules, yaks, donkeys and oxen. The oxen, which Cawdor called 'the Express Dairy corps', were outstandingly unreliable. When they grew tired they simply sat down. The ponies on the other hand spent most of their time bolting through the scrub, kicking off their loads and scattering a trail of wreckage. And it was not only the baggage which suffered.

Cawdor wrote on 14 April: 'W took out his notebook and for some reason his pony — which previously had refused utterly to budge — became galvanised into life and bolted. It charged straight into my beast and W who was reaching for the reins was caused by the collision to cut a voluntary — I grabbed the pony and picked up W who was fortunately none the worse for having fallen on "stoney ground".'

The journey did not start well. K-W was used to delays and upsets caused by pack animals playing up and the lack of urgency so often displayed by their handlers, but bad weather and the unruly animals upset Cawdor. When, on 15 April, his pony's harness broke during a dust storm, the young man turned on K-W: 'I cursed W — quite unreasonably — for not stopping to give me a hand, and we had a row, made it up and felt atmosphere lift immediately — As he says if two chaps like us, who don't know each other very well, go into the blue for a year, they are bound to have words until they settle down — I'm thinking he is a better philosopher than I.'

There were other rows to follow. Four months later, at the

beginning of August, Cawdor became intensely frustrated by what he perceived as the slow pace of the expedition, finding it hard to appreciate that botanising and plant collecting cannot be rushed.

At Lunang he wrote this entry for 2 August:

> It was no use waiting for W for the last time I had seen him, he was just beginning to come down the first scree from the top — I was then at the head of the second ledge where the road dropped in the rhododendron forest — he was half an hour's walk behind then — move at his pace — God knows how he does it — He got 20 minutes start on me this morning and in climbing slowly to the summit in $1\frac{1}{4}$ hours left him out of sight — There's not much companionship to be got out of such a chap — It drives me clean daft to walk behind him — Stopping every 10 yards and hardly moving in between — In the whole of my life, I've never seen such an incredibly slow mover — If ever I travel again I'll make damned sure its not with a botanist. They are always stopping to gape at weeds.

He goes on to say how he wishes he was walking in Tibet with an old friend of his from Scotland, and then continues: 'As it is I'm bound for Lord knows how long to wander this damned country with a man who can only shuffle along like a paralytic — I could forgive W most things if only he could walk — though, evidently, God never intended him to be a companion to anyone.'

Clearly the resentment festered because in the entry in K-W's diary for 3 August, he writes: 'I irritated Cawdor, who had had a bad night and was feeling unwell from the effects of the last three days; and he went straight of the deep end. Smoothed him down.' But despite the difference in age and personality, Cawdor clearly did admire K-W, though they never became close friends. Certainly there were times when the plant hunter caused him great amusement: '25.4.24: W wears long pants with shorts — it is admirable — I shouldn't have the courage to do it.' Cawdor normally wore riding breeches and puttees.

But he was cross with K-W again when they ran out of curry powder. K-W had not included it in the stores bought in Britain, and had failed to get any in India. What they had was obtained in Kalimpong.

'It is a sad loss,' Cawdor wrote, 'as we depend on curry as a staple food. We are badly off in the way of victuals now, being reduced to dahl, rice, eggs, *tsamba* and tea. We want to keep our stores as long as possible — we hope to be able to find some capsicum with which we may be able to make up some sort of curry powder.'

It was ironic that he should complain, for of all the Kingdon-Ward expeditions to date this was probably the best stocked, with stores for a varied diet. Indeed the shopping list from Fortnum and Mason reads like a tuck box that would have sent Billy Bunter into paroxysms of delight. Fortnum and Mason rather charmingly described it as Captain Ward's hamper. Included in it was Nestlé's milk, blackberry and apple, gooseberry, apricot and strawberry jam. There was tinned Dorset butter, Quaker oats, Heinz baked beans, mincemeat, HP sauce, Yorkshire relish, blocks of 'hors d'oeuvre' (which Fortnums now believe were blocks of foie gras), dark Mexican eating chocolate, tins of *café au lait*, ground coffee, Fortnum and Mason cocoa, soup squares, a smoked York ham, saccharine, tea, Wrigley's chewing gum, and herrings in tomato sauce.

Of course the food did not always appear on the folding Abyssinian camp table in a conventional form. On 29 May mince pies were ordered for dinner. K-W's diary entry remarks: 'Dick, our cook, covered himself with glory by making mince pies out of a tin of sardines, instead of mincemeat. They weren't at all bad, though rather unexpected.' Cawdor gives a fuller account:

> We got to our house at 18.45 having come down the best part of 4000 ft in an hour. We dispensed with tea and had supper whenever it was ready — Soup, curry (our staple and plenty of it) and tea. We were dying for drink, not having had anything since breakfast — We had given a tin of mince meat to Dick with instructions to make something out of flour and eggs to do for pies — the pies turned up all right, but he had opened the wrong tins and stuffed

them with sardines. W opened one and said, 'Good God there's bones in this mince pie!' – the sardine patties were excellent, if unexpected.

It is a common feature of all long expeditions that those on them become quite obsessed by food, especially as the local diet in remote places tends to be monotonous in the extreme. K-W wrote in his journal:

> Ordering meals is for us quite simple. 'What will you have for breakfast?' says Dick. 'What is there?' I ask. 'There is rice, and dahl and *tsamba*, and eggs.' 'Oh. Scrambled eggs, rice pudding and some *tsamba*.' 'What for dinner, sir?' 'Dahl, curry and scrambled eggs.' Sometimes there is a fowl, and we have roast fowl, scrambled eggs and rice pudding for dinner. Rice pudding, scrambled eggs and cold fowl for breakfast. Sometimes we have a jam week, or a biscuit week. In Tibet there is hair in everything – in the butter, the milk, *tsamba*, even the eggs.

In June they were unable to get fresh milk, 'which is sad', Cawdor notes,

> since it precludes buttered eggs from our very limited menu – It also cuts out rice pudding, which I believe I could live on by itself – unfortunately W doesn't seem to care for it much.
> We dined rather miserably off an incredibly tough fowl. It was such hard work to eat it that one was hungrier at the end than at the start – It had been carried up the pass in a basket on an old woman's back, together with the eggs and, I suspect, the onions. I thought it was a muscular bird, for it escaped in camp yesterday and legged it up the hill like hell with all the boys after it – We also ate poached eggs and stewed figs, but the backbone of our meal was composed of chapatties – By God I could do justice to a damned good slab of Figgy Duff tonight.

To the Land of the Great Gorge

K-W found Cawdor's insistence on having a bath whenever possible almost eccentric. In one journal entry Cawdor wrote: 'I feel cleaner too after my tub last night – W thinks the weather too cold for such business, but I shall try to have a bath at least once a week whilst I'm here. It makes all the difference.' Even at home in England K-W believed that too much bathing was quite wrong. He would light-heartedly lecture Florinda, who took two hot baths a day, on his theory that hot baths and effeteness had brought about the downfall of the Roman Empire.

Another thing K-W and Cawdor had in common, and something that seems to affect most travellers who are separated from home and civilisation for long periods, were fits of deep depression. Usually they were linked to bouts of ill-health. During May Cawdor suffered from a persistent stomach upset. On a gloomy Sunday at Tsela Dzong he sat down and wrote:

> I think the chief reasons for my present discontentment are firstly my sickness which completely kills what pleasure there is in life, secondly my lack of knowledge of how to do justice to the ethnological and geological opportunities available (in other words I realise that I am a failure!), and thirdly the feeling of being utterly cut off from the ones one cares for.

It is possible that K-W was partly to blame for Cawdor's sense of failure. When he travelled with companions he tended to let them get on with their job. He assumed they knew what they were doing and would not interfere, or offer as much of his wisdom, experience and expertise as he might have done. Ronald Kaulback, who travelled with him some years later, made exactly this point. At any rate it was clear that Cawdor did feel hard done by, and grumbled quietly into his diary.

> 13.5.24: Tsela Dzong: After supper I lit the red lamp and endeavoured to put some [photographic] plates in my quarter-plate tank – It was fairly dark when I started, but the boys in the kitchen next door suddenly stoked up the fire like hell and the light streamed through the half-inch

> gaps in the planking. However, I think I managed to avoid this — What really wrecked everything was W choosing to walk down the passage with his electric searchlight — As I am separated from the passage by another flimsy wall huge beams broke in upon me and I could do nothing to prevent it.

Next day he examined his work in daylight.

> Three out of six of my plates have come out pretty well including one of the rhodos. The other three are badly streaked by the brilliancy of the searchlight — I was a mug to try to do any photographic work in such a rotten room, but I thought I should be giving W less trouble if I did it here — In future I shall do as much as possible in my bed — It is the only light tight place.

Cawdor would have been the first to admit upon reflection that in fact his real or imagined misfortunes had nothing to do with K–W, but nevertheless he continued to scribble out his misery in indelible pencil:

> There is only one experienced traveller with whom I should like to go anywhere and that is Louis Clark — the one thing that makes it difficult for me on this trip is not knowing any Hindustani [the language of their servants] — I have to rely so much on W.
>
> One is, I suppose, bound to feel homesick at times and long for one's friends, more especially when one is sick physically as well — I have noticed before when travelling that there are always periods of great mental depression.

He was still feeling pretty low the following morning and ready to grumble: 'I turned out about 0700 and cut up and put away negatives. W still slept soundly at 0815, so God knows when we shall get any breakfast.' K-W was more charitable towards Cawdor, however. In a letter to Hinks in January 1925, he said: 'Cawdor has been a delightful companion.'

Being such a seasoned traveller, K-W was able to suppress

the inevitable homesickness that he felt from time to time, but even he could not disguise the sense of excitement as an expedition drew to a close and home beckoned. In *The Riddle of the Tsangpo Gorges*, which tells the story of the expedition, he writes:

> No one who has never wandered far out of his orbit can conceive the tremendous pull exerted by home as he approaches nearer and nearer the centre of his system. We are deeply under the influence of the major body, being but two marches from India, and fatigue vanished; we could have marched – nay, felt compelled to march – all day and all night, drawn irresistibly towards the sun of our civilisation.

Although Cawdor found it hard to establish a close friendship with K-W, he was most certainly concerned for his well-being. On 28 May, when they were at Tsela Dzong, they were passing through a narrow gap after finding a group of *Primula pulchelloides* just coming into flower, and bushes of *Rhododendron triflorum*. K-W made three separate seed collections of the latter, which included a suspected variety with flowers which were either pure yellow, or flushed with ochre, mahogany, or almost salmon-pink. To collect seed it is often necessary to scramble right into a bush, and on this occasion he was struck in the eye by a whippy branch which bruised his eyeball. By the time they got back to their base he was in terrible pain. That night Cawdor wrote: 'W's eye is very painful, he seems to have bruised the eyeball – I gave him some boracic to make a lotion in warm water – I also gave him a morphia tablet as he asked for it and he turned in about 1200.'

The morphia helped K-W to have a good night's sleep, and he was fit for work the following day. Cawdor seemed rather put out that K-W had made such a swift recovery:

> W had a good night, slept like a log with the morphia – I wasn't ready when he started immediately after breakfast, and had a good many things to do at home, so I did not accompany him – No doubt one misses a lot through not being prepared to go off for a whole day as soon as one

has eaten one's breakfast, but there are few luxuries in this life and one I think worth observing is ten minutes, at least, in which to digest a meal, however indifferent it might have been.

Cawdor controlled the medicine chest, which contained the morphia tablets. It was a drug he disliked, and he was alarmed that K-W was ready to use it so freely.

In his journal entry at Gyala on 20 July, he wrote:

> W is suffering from an aching tooth. He asked me for a morphia tablet, which I gave him ... He seems to have a different idea of the uses of the drug to what I have. I rather hesitated in accepting Hislop's offer of them at Gyantse, but thought they might be useful in the event of an emergency — I have always regarded them as things only to be used for getting one night's sleep in six; for killing severe pain without waiting for the doctor's arrival, or for enabling one to die with the minimum of discomfort — W appears to regard them as a tonic ...
>
> Toothache is bad and I am sorry for W, but I hope he won't always be asking for morphia — it is difficult to refuse when one knows a man's suffering. But it is easy to form a habit and with my extremely slender knowledge of medicine I am afraid of giving too much. I almost wish I hadn't got the stuff with me.

In fact Cawdor's teeth were in a bad way too. One had broken and the filling had come out of another. As he wrote in his journal: 'My tooth hasn't been quite so bad today though I have suffered from a dull ache all the time — It's sure to be an abscess judging by the lump on my jaw and the smell of the wads of cotton wool I take out of the tooth — I've smelt nought like it since W and I got leeward of the Madras main drain.' Eventually they were both able to control their dental problems with plugs of raw opium.

Although they suffered all the usual maladies of such an expedition — stomach upsets, fever, flea and louse bites, toothache and altitude sickness — they were never seriously ill. There was however one bizarre attempt on their lives, on New Year's Eve. Cawdor recorded the event in his usual laconic style:

> When we were having supper outside our tent by the fire, a woman — one of the coolies — was led up like a lamb to the sacrifice — her hands bound by a thong. She was charged with mixing mud with our flour — the chupatties have certainly tasted a bit gritty and we have both been complaining of the Tongjuk flour — the Dzong Pon sold the flour and we cannot discover what motives the woman had — unless she is a born murdress — She was led off cursing and revilling shrilly and we did not pursue the case to any length. This morning we ordered the flour to be thrown away and decided to make shift on *tsamba* — this was the first attempt on our lives — Although it was not as bad as it might have been, one cannot get much pleasure from eating muddy chupatties.
>
> We have both been suffering from bellyache for the past two days — No doubt due to the diet of sand provided by this infernal bitch — A suitable punishment for her, it seems to me, would be to make her eat the flour she doctored.

As well as surveying and map making, Cawdor was also in charge of the coolies and transport, a job that K-W always gave to his travelling companions because he found the day-to-day organisation of a camp or a march irksome and time-consuming. And so Cawdor became, in his own words, a 'sort of head transport wallah, Quartermaster, Good Samaritan, and Pioneer Sergeant'.

He had a direct if unsubtle way of dealing with the men. Early one morning he and K-W were woken by a man beating on a suspended wooden beam which was used as a gong. ''W threw a boot at the enthusiastic beater but missed him — I turned out and registered with a boot — Furthermore, I got out my dirk and cut down the beam.' He threw it in the river: 'It made a most consoling splash!'

It is understandable that tempers wore thin from time to time, for K-W and Cawdor were travelling through some of the most difficult and hostile country in the world, from the cruelly bleak uplands of Tibet, where grit-laden gales sand-blasted any exposed parts of the body, to the mountain slopes and gorges, swept by rain and sleet or blanketed with thick, soaking mists. In *The Riddle of the Tsangpo Gorges*, K-W gives a clear idea of the discomforts when he describes Pemako, a province of Tibet on the south side of the Himalaya: '[imagine] a climate which varies from sub-tropical to arctic, the only thing common to the whole region being perpetual rain; snakes and wild animals, giant stinging nettles and myriads of biting and blood-sucking ticks, hornets, flies and leeches, and you have some idea of what the traveller has to contend with.'

While the area was rich in flora, there were long treks between collecting grounds. On the high plateau the plants were sub-arctic, and therefore of no garden value, and in the sub-tropical rain forest the beautiful rhododendrons were too tender for open gardens and could only be grown under glass. It was the temperate forest of the gorge country, and the flower-strewn alpine meadows that yielded four-fifths of K-W's horticultural harvest: superb hardy rhododendrons, from dwarfs that carpeted the ground or flowed over the rocks to shrubs and trees; roses and berberis; wonderful primulas and meconopsis. But because K-W was making herbarium collections of dried specimens that would go to the Botany Department of the Natural History Museum and to the Royal Botanic Gardens at Kew, he had to collect virtually everything, whether it was likely to be suitable for gardens or not.

One collecting commission he received in Tibet itself. He was asked to gather seeds for the then Dalai Lama, who was a keen gardener and grew a large number of plants at his private residence, Norpu Lingka (Jewel Park), on the outskirts of Lhasa. K-W sent him nearly forty species of primula and meconopsis 'and other showy plants which could be easily raised'.

Horticulturally the expedition was an outstanding triumph, best remembered for the introduction of the Tibetan blue poppy, which became known as *Meconopsis baileyi* but which has since passed through various taxonomic transformations from *Meconopsis betonicifolia* var. *baileyi* to the now plain *Meconopsis beton-*

icifolia. K-W originally named it in honour of Lt-Col F. M. Bailey, a man he greatly admired.

Bailey, who was Political Officer for Sikkim, was a skilful and inspired traveller and explorer, as well as a talented gardener. Although he himself had failed to solve the mystery of the Tsangpo gorges and waterfalls, when he heard that K-W was following in his footsteps he was generous with his advice. Writing from Gyamda Dzong, to Hinks at the RGS, K-W said: 'We travelled by the high road from Kalimpong to Gyantse, staying two days with Bailey en route. He was awfully good to us – gave us final instructions, information and advice and away we went with his blessing.'

Being such an experienced traveller himself, Bailey knew the value of a little light relief in the face of the hardships of the wilds. In October 1924 he sent Cawdor a cutting from *The Times* reporting that the Cawdor family chauffeur, William North, had been fined £5, with three guineas' costs, at Marylebone Magistrates' Court for driving at between 25 and 30 m.p.h. down the Marylebone Road, skidding and practically flattening a constable on traffic duty.

It was during his journey to the Tsangpo gorges in 1922 that Bailey had picked a specimen of the blue poppy at Lunang, Kongbo, in south-eastern Tibet, and had pressed the specimen in his pocket book. By the time it reached England it was in a fairly damaged condition, but interesting enough to arouse a good deal of excitement.

In fact Bailey was not the original discoverer of the plant, as Sir George Taylor explains in his classic work, *An Account of the Genus Meconopsis*. He credits its discovery to the Catholic missionary priest and naturalist, Père Delavay, who found it growing in north-western Yunnan in 1886. K-W's important contribution was to collect viable seed in Tibet. It proved to be a superb garden plant.

When he spotted it blooming among bushes two hundred miles east of Lhasa, he first mistook it for a gorgeously plumaged bird. Before it had proved itself in British gardens, he wrote this of it in *The Riddle of the Tsangpo Gorges*:

Beautiful as were the meadows of the rong, a patchwork of colour exhaling fragrance, nevertheless the finest flowers hid themselves modestly under the bushes, along the banks of the stream. Here among the spiteful thickets of hippophae, barberry and rose, grew that lovely poppy-wort, *Meconopsis baileyi*, the woodland blue poppy. This fine plant grows in clumps, half a dozen leafy stems rising from the perennial rootstock to a height of 4 feet. The flowers flutter out from amongst the sea-green leaves like blue-and-gold butterflies; each is borne singly on a pedicle, the plant carrying half a dozen nodding, incredibly blue four-petalled flowers, with a wad of golden anthers in the centre. The foliage is startling enough, the lower stalked leaves reaching a length of 2 feet, the upper ones sessile, their round-eared bases clasping the stem. Never have I seen a blue poppy which held out such high hopes of being hardy, and of easy cultivation in Britain. Being a woodland plant it will suffer less from the tricks of our uncertain climate; coming from a moderate elevation, it is accustomed to that featureless average of weather which we know so well how to provide it with; and being perennial, it will not exasperate gardeners. If it comes easily from self-sown seed, as a few species do, it will be perfect.

This passage, more than any other, demonstrates why K-W was such an outstanding plant collector; he really did understand what was needed in a good garden plant.

Seventy-five per cent of the seed he sent home in February 1925 germinated in the fifty gardens it had been sent to for trial in England, Scotland and Wales. At the Royal Horticultural Society's June show in 1926 it was awarded an unchallenged Award of Merit, the following year it received an RHS First Class Certificate, and the year after a Gold Medal in Ghent. In 1930 it was featured in the *Botanical Magazine.*

The plant was a sensation, and the mass plantings of it in Hyde Park in London, and Ibrox Park, Glasgow, drew admiring crowds. At the 1927 Chelsea Flower Show tiny seedlings were sold at a guinea apiece, but five years later it had become so widespread that packets of seed were advertised at a shilling. But despite its popularity and undoubted beauty, K-W did not in the end rate it as a first-class garden plant because of its short flowering period.

While the blue poppy grabbed all the headlines, the most important part of his collection were the rhododendrons – ninety-seven species discovered on this expedition alone, a score which was only beaten in 1931 when K-W collected 106. Among the Tibetan rhododendrons was the now famous 'Orange Bill', a jaunty reference to Prince William of Orange.

K-W described how he found it in *The Romance of Gardening*:

> There was snow everywhere, snow and tumbling water. Far below, the valley, hemmed between gaunt granite cliffs, broadened out a little, and vast avalanches of snow, forty or fifty feet deep, blocked the way; beyond lay the outposts of the forest, stunted fir trees straggling out beyond the main body. And half way down towards the fringes of the forest, in the dense scrub, I caught a glimpse of orange, vivid amongst the crimson glory of the Rhododendrons; a billow of foliage shone blue-green.
>
> It was raining steadily now, and the task of forcing a way through the shoulder-high scrub was no light one. However, there was the bush, not two hundred yards up the slope, and I butted into the thicket, forcing my way through. It was a stiff fight, but at last, out of breath and out of temper, torn, soaked to the skin, and cold, I reached my goal. It was that dreamed of, but scarcely hoped for, treasure, a real orange-flowered Rhododendron! And in that moment of triumph I was almost delirious with pride and joy. 'Orange Bill – the Prince of Orange Rhododendrons,' I cried aloud; for I felt lyrical. There was only one bush that I could see, and it bore few flowers. I hardly dared cut more than one spray for the herbarium; I should need it all for seed.

That evening in camp he made this field note: 'KW 5,874. Rhod. Sp. (Roylei). Doshong La. 12,000–13,000 ft. 29/6/24. Flowers orange. Bush of 6 ft. Growing on steep rocky slopes amongst dense Rhododendron scrub. Leaf buds not yet broken. Foliage bright glaucous, visible afar; a most striking bush.'

The final sentence to the note was added later when he went back in October to harvest the seed, and the remarkable colour of the foliage struck him more forcibly. By then the new growth had fully expanded and 'shone more brightly than in June, a wonderful blue-green like verdigris, it stood out violently against the white background'. In October the snow had fallen.

It had been something of a battle to get its flowers, but to reach the shrub to collect capsules of seed was even more difficult.

> To say that I progressed two yards in five minutes up that accursed slope would convey no inkling of the struggle. It was a fight all the way, through frozen tangle. To go straight up was impossible. I dodged which way I could. After pushing in vain against an unyielding barrier, I would seek passage to right or left, at the same time trying to keep my eye on the prize. Yet so steep was the face, that I as often lost sight of the goal, struggling up to my neck in the thicket. Heaving against the stiff upsweeping branches which fended me off breast high, I finally overbalanced and fell back against the more yielding bushes behind; which gave way, so that I broke through the tangle and collapsed amongst snow and vegetation, half supported, in a helpless position. All the while my hands were being lacerated by the steely branches. After about an hour of this sort of thing, suddenly the bush loomed up beside me like a great puff of blue-green smoke, and reaching out, I grabbed half a dozen capsules. Then I searched the bush, but it had flowered poorly, and I got little more from it; but floundering about on the cliff, I presently found several smaller bushes – the verdigris-coloured foliage was unmistakable, though I had not seen

these bushes in bloom — and in all I secured perhaps a score of the fat little capsules, each with its hundreds of seeds.

The harvest was enough to bring a fine plant to British gardens, but it was not a new species, turning out to be a particularly fine form of *Rhododendron roylei*. It took him two attempts in appalling weather conditions to complete the collection, and he later commented: '... it is a relief to think that the seeds are germinating considering the awful strain on my temper while struggling in the accursed cold muddle.'

There is a mystery over one rhododendron which Lord Cawdor spotted at Temo La. He wrote in his diary on 8 June 1924:

> Close to the spot where I sat and read Walton, *The Compleat Angler*, I thought I saw a strange rhodo — I said to W 'Is that a new rhodo or is it a variation of *thomsonii*?' — W looked and said it would bear investigation — Thus was *R. cawdorensis* discovered. It proved to be quite different from *R. thomsonii* and had cream-coloured flowers with purple markings and a pretty habit of growth — we marked the tree and went up the flume keeping along the edge of the very precipitous Khad — down below we could see quite a number of meadows, but it was too deep to make a descent — We went back to the big camping ground and got down to some boggy fields where we found some more plants of 'my rhodo' — We found a pinker variety — Finally we got down to the burn where we discovered two more new rhodos. One a pink one with dark leaves and furry undersides and one very pretty pure 'white of egg' rhodo.

The mystery is that there is no mention of *R. cawdorensis* by name in *The Rhododendron Handbook*, the stud book of rhododendron growers, or in K-W's field notes on the rhododendrons he collected during the expedition. Did K-W dub it *R. cawdorensis* there and then in honour of its discoverer, did Cawdor do the naming himself, or was the name disallowed later? The latter is most likely. It was certainly not familiar to K-W. He numbered it

KW 5,759, and his field notes, which coincide with Cawdor's record of the discovery, read:

> Temo La. 13,000 ft. 8/6/24. Flowers white, cream, pink, rose or crimson, more or less spotted with purplish crimson. Seedlings have no underleaf indumentum, but a scurf of brown hairs on undersurface. These may remain white, or slough off, but usually turn rusty red. In the forest, a shrub of 10–15 ft. Forms a continuous dense scrub 4–5 ft high above the tree line. It keeps to the open where the large *lacteum* buries itself in the forest.

In the stud book it is named as *Rhododendron agglutianum.*

K-W revelled in the masses of rhododendrons, thrilled by the rich pattern of colour, and by the fact that he had discovered species quite new to cultivation.

He loved to make up popular names for his discoveries, too. Thus there was 'Orange Bill' and 'Coals of Fire' (*Rhododendron cerasinum*), with its huge, drooping trusses of fleshy blazing scarlet flowers, and five coal-black glands the size of shirt buttons at the base of each corolla. 'Yellow Peril' (a *Rhododendron campylocarpum* variety) was named for its 'aggressive abundance'. 'Scarlet Runner' (*Rhododendron repens* var. *chamaedoxa*) was found creeping over the rocks. He wrote in *The Riddle of the Tsangpo Gorges*: 'For a moment we just stared at it, drunk with wonder. It lay absolutely flat on the rocks, no part of the plant, not even the corolla, which is considerably larger than the leaf, rising two inches above the surface; stem, leaves and flowers cling as closely as possible to the ground.'

'Plum Warner' (*Rhododendron campylogynum*) had 'absurd little plummy mouths pouting discontentedly at us'. There was 'Plum Glaucum' (an *R. tsangpoense* variety) with its dark cerise flowers; the intense crimson of 'Scarlet Pimpernel' (an *R. repens* variety), growing with several other species, 'woven into dense chromatic carpets'; the racy carmine of 'Carmelita' (*Rhododendron* var. *chamaethauma*), and the fragrant white flowers of the 'Madonna' (an *R. nuttalli* variety). All but 'Madonna' were collected at the Doshong-la, which was an extraordinary natural rhododendron garden. K-W wrote:

> It is impossible to do justice to the rhododendrons at the Doshong-la as we saw them in June; the valley, flanked by grey cliffs, roofed by grey skies, with the white snowfields above, spouting water which splashed and gurgled in a dozen babbling becks; and everywhere the rocks swamped under a tidal wave of tense colours which gleam and glow in leagues of breaking light. 'Pimpernel' whose fiery curtains hang from every rock; 'Carmelita' forming pools of incandescent lava. 'Yellow Peril' heaving up against the floor of the cliff in choppy sulphur seas breaking from a long low surf of pink *lacteum*, whose bronzed leaves glimmer faintly like sea-tarnished metal.

And it was not only the rhododendrons that spread this breathtaking tapestry of fragrant colour throughout the meadows and valleys of the Doshong-la. The primulas clamoured for space in the grass, by the snow-swollen brooks, and among the shrubs.

There was *Primula alpicola*, which he called 'Joseph's Sikkimensis Primula', because its five colours — violet, purple, maroon, sulphur and cream — reminded him of Joseph's coat of many colours. The sulphur and cream varieties formed, he said, 'an herbaceous border along the forest paths'. But there was one primula that stood out from among all the rest, the giant cowslip *sikkimensis*, which he named *Primula florindae*, for Florinda. It transcended the beauty of the 'Moonlight' primula (*Primula sikkimensis*). K-W found it growing in profusion on both sides of the Tsangpo:

> It was most happy in the woodland brooks, which in summer overflow and flood the thorn brake. Here it manned the banks in thousands and, wading into the stream, held up the current. It choked up ditches and roofed the steepest mud slides with its great marsh-marigold leaves; then in July came a forest of masts, which spilled out a shower of golden drops, till the tide of scent spread

and filled the woodland and flowed into the meadow to mingle with that of the 'Moonlight' primula. And all through August it kept on unfurling flowers and still more flowers, till the rains began to slacken and the brooks crept back to their beds, and the waters under the thorn brake subsided ...

This stupendous primula grew to a height of four feet. K-W also named a dwarf golden meconopsis after Florinda. He collected seed, but unlike *Primula florindae*, it failed to establish itself in British gardens.

He was surrounded by such an abundance of beauty that he considered excluding some plants from his garden collecting. But in the end it was impossible to ignore those which his eye, instinct and judgement told him would be good. He wrote in *The Riddle of the Tsangpo Gorges*:

> There was a barberry. Now barberries in Britain, and particularly Asiatic barberries, are legion, and though most of them are as like as two peas, they have all been proclaimed swans; the best way to avoid trouble was to avoid barberries, cotoneasters, viburnums, and other collectors' geese altogether. Therefore, despite the fine glaucous foliage of this bush, I turned away unimpressed. But when the fruits ripened to coral beads dangling from the jet-black stems and the blue-green foliage was shot with scarlet and old-gold in the fall, I could no longer refrain from collecting seed.

By the beginning of August the first stage of the botanical expedition was almost at an end. Writing to Hinks from Gyanda at the beginning of September, he declared with evident satisfaction: 'We have been very successful, with nearly 50 species of rhododendron, about 40 primulas and 10 poppies (meconopsis) including what I believe is the first dwarf yellow meconopsis known (*Meconopsis florindae*). Most of these plants are new, many of them are of great beauty, and we hope to secure seeds of them all.'

October was the month of the seed harvest, and it proved to be gruelling hard labour. The weather was cruel, bitterly cold with heavy falls of snow. In the alpine meadows he had to dig with his hands in the snow to find the ripe seed capsules of many of the plants that he had seen only a few months earlier in a glorious display of colour. But he was able to locate everything he wanted and more besides.

The effects of strenuous marches, indifferent diet and long hours of work were beginning to tell, as an entry in his journal attests:

> How different to the early months. In June the days are not long enough. Up at daylight, and working till late at night. The pace is too hot. By July one is tiring. By August one is worn out. It is necessary to take it easy for a bit. No longer does one get up at dawn, the edge of keenness is blunted. And now in late autumn the work recommences, harder than ever, and finds a tired man. One must call up all one's strength for the winter campaign; the hardest task yet lies before one.

After drying and packing the seed harvest and the herbarium specimens, some two hundred species in all, K-W and Cawdor prepared for what K-W called the 'big push', the trek to the fifty unexplored miles of the Tsangpo gorges where the river cuts its way through the mountains. 'We had been looking forward to this all through the summer,' he wrote to Hinks at the RGS.

> Twenty-four porters were at last secured, and this levy almost emptied the villages on which we had permission to draw. As it was we had to put up with half of them women. Everything was cut down so that we could carry the necessary food and instruments. Two thirds of the porters carried rations, and beyond the clothes we stood up in and our bedding, we carried only surveying, photographic and plant and seed collecting outfit.
>
> Everything was in our favour. We had a guide who five years before had been through the gorge; porters and

food for twenty days; the water low; and a fairly good season — not the best, but quite good. So I think we should have been rather mugs if we had not succeeded.

On 16 November 1924 they left Gyala to the sound of prayer drums and the aroma of incense. It was the biggest caravan of the expedition; in addition to their twenty-four Monbo porters there were two personal servants, two dogs, two fowls, and a sheep whom Cawdor named Homeless Horace. At one stage they had to climb part of a cliff using log ladders, and Cawdor noted: 'I hauled Homeless Horace the sheep up with a rope while an old lady beat him with a stick from below — His ascent can only be described as extremely reluctant.'

K-W did not waste any time on the journey and continued his search for seeds, something which Cawdor found astonishing. His entry for 19 November reads: 'W caught up and passed us here very busy looking for new rhodos — It is a wonderful revelation — after eight months in this infernal country I shouldn't have imagined anyone would wish to see another rhodo again — I am damned sure I don't!'

The first part of the big push took them through swamps and dense forest where the undergrowth grew six feet tall. But closer to the gorges the country became incredibly rugged, with dangerous cliff faces to negotiate. The women porters had to be hauled up or lowered on ropes, and faced with one particularly difficult descent to the river, they simply sat down and wept.

By 28 November they had reached the narrowest part of the gorges, and were virtually trapped between cliffs rearing hundreds of feet high on either side. They made a camp where the river took a great bend to the north-west, and K-W wrote to Hinks: 'Exploring round the bend from our bivouack that evening, I turned the corner and saw the gorge filled with spray, and the river dropping vertically over a ledge. "The falls at last," thought I. However, it wasn't. We couldn't get within six hundred yards of the fall, but estimated a vertical drop of about forty feet from the spray and noise and the level of the river below.' He named it the Rainbow Fall.

To the Land of the Great Gorge

The following day was a nightmare as they scrambled and hauled themselves up the cliff face to spend a miserable night in pouring rain on the hillside. For the next few days they followed the winding course of the gorge, getting occasional glimpses of the river. At one point, when they were able to get down to the river bed, they found it had dropped 2,230 feet from the point at the Rainbow Fall where they had measured the altitude.

They then followed the river for two miles before again being hemmed in by sheer cliffs, their way blocked by an enormous spur. Their porters claimed there was no way to get beyond the spur, but at dusk that evening, when K-W was searching for a spot to set up the theodolite to take readings for their survey of the country, he spotted a good path leading over the spur. He also spotted a camp fire on the far side of a valley formed by two spurs, which the porters insisted was inaccessible. Cawdor followed the path and found a perfect camping spot. He wrote: '... after that the opposition [the porters] collapsed and agreed to take us. So we set out on a final five days' attempt.'

The weather closed in again with blinding snowstorms, and it took two days to reach the top of the spur, but when they did they were rewarded by the sight of a long stretch of river four thousand feet below. They descended two thousand feet and spent a wretched night in wet, bitterly cold conditions on a precipitous hillside.

Cawdor was unwell, having suffered for days from toothache and lack of sleep. '... it was wonderful how he kept on; he got little rest, poor fellow – we either had to go on all the time, or give the whole thing up,' K-W recorded. Leaving Cawdor at the camp, K-W scrambled down the remaining two thousand feet to the river. At this point he found that it fell about one hundred feet in a quarter of a mile, creating a series of small waterfalls until it reached a ledge forty feet high which was breached by a narrow gap through which the river raced. Judging by a deep, calm pool scooped out of the river bed, he concluded that when the snow melted the torrent would create an impressive, if temporary, fall over the ledge, but this was not the deep gorge of the fabled falls. He named the ledge Takin Fall, because while he was exploring it his porters had shot two takin, a variety of wild sheep.

The Tsangpo enchanted K-W and he wrote about it again and again. In *The Riddle of the Tsangpo Gorges* he describes it from the camp two thousand feet above it:

> ... when you think the gap must be sealed up, and the door bolted and barred, out of the very heart of this tomb, swinging round the spurs, leaping the rocks, comes the Tsangpo just as hard as it can go, a roaring, bouncing, bellowing flood. You see one flash of green, like jade, where the sunlight gleams on a pool far up the gorge, and after that all is white foam.

The river would disappear as suddenly as it appeared. Still viewing it from the camp site, he wrote:

> Immediately below this point, the boulder beach comes to an end at the front of the cliff; and what happened next we could only guess, for the river, after hurling itself through the gap, rushes headlong into a gorge so deep and narrow one can hardly see any sky overhead; then it disappeared.

By climbing a cliff he was able to get a further and final glimpse of the Tsangpo:

> Below the whirlpool created by the first fall, the river flowed smoothly for about a hundred yards and was a dark jade-green colour; here it was not more than thirty yards wide, and must have been incredibly deep. Flowing more swiftly, it suddenly poured over another ledge, falling in a sleek wave about forty feet. Scarcely had the river regained its tranquilly green colour, than it boiled over once more, and was lost to view round the corner.

They were denied the chance to prove or disprove the legend of the falls for certain because it was impossible to enter the five miles of gorges where there were supposed to be seventy-five falls guarded by individual spirits, and there was no vantage point they could reach for a view from above. But according to calculations based on the measurements they took, K-W and Cawdor concluded that if there were any falls in the deep, sunless gorges, they would be no more than about twenty feet high.

Having explored the Tsangpo gorges as far as they could, they now turned their attention to a range of snow peaks.

On 28 December they set off in glorious weather to survey the mountains, but before long the weather turned and they were trapped in a snowstorm that raged for two days. Finally they had to abandon their surveying work and make for a pass that would take them out of this arctic wilderness. The hard snow granules, whipped up by a high wind, slashed and scoured their faces until the skin was red and raw; their eyes were bloodshot and their noses frost-bitten. At last they crossed the pass to encounter one of the extraordinary phenomena of the Himalaya. At their backs lay a deep and bitter winter, but before them was benevolent spring.

> Behind us the dead plateau [wrote K-W], wrapped in its dazzling white shroud, stretched out its frozen limbs to the pale porcelain mountains, all frothy with cloud. Except for the moaning wind and the swish of the driven snow blast, complete silence reigned. There was no tinkle of water, no song of birds, not even the flutter of a leaf; everything was dead, or fast asleep, or gone abroad for the winter.
>
> In front of us, the mountain dropped away steeply to the valley, the snow stopped abruptly, and the dark mysterious forests of the southern slopes of the Himalaya began.

They had not been able to penetrate the deepest recesses of the Tsangpo gorges, and the weather had prevented them from measuring the mountains, but they had discovered more about the Tsangpo River. Cawdor had done some useful surveying and map making, and K-W had reaped a rich harvest of fine new plants for British gardens. In January 1925 he wrote to Hinks: 'If we had had complete and extravagant success — were such a thing possible — we shouldn't be any happier.' Still recovering from the hardship he had endured, he declared his intention of returning to the gorge country to explore even further. K-W and Cawdor were tired, physically reduced by poor food and illness, but content in the way only explorers can be when they have completed a long and arduous journey that has ended well.

After a series of minor hold-ups they reached Rangyia on one of the small tributaries of the Brahmaputra in Assam. It was a little up-country railway station, simple to the point of being spartan, but for them it was luxury. They made up their beds on the stone floor of the waiting room, and after strolling up and down the platform for a while, went into the refreshment room for the exquisite pleasure of being served by white-robed attendants. It could have been the dining room of a fashionable London hotel, so great was the contrast with the harshness of their existence over the past months.

'How we enjoyed that meal,' wrote K-W, 'seated at a table with a clean white cloth. And how the ordinary Englishman, travelling in India, would have turned up his nose at the curried chicken. But to us it was manna.'

The next day they boarded the mail train for Calcutta, where they had started their journey nearly twelve months before.

6

Kingdon-Ward Territory

IN 1926 K-W set out on a short expedition to the northern frontier of Burma, down the Lohit River, and through the Mishmi Hills in Assam. It was, in a way, a trail-blazer for a much more ambitious project planned for 1927 and 1928. He explained in the introduction to *Plant Hunting on the Edge of the World* the purpose of the 1926 trip, and the one to follow:

> (i) to collect seeds of beautiful hardy plants for English gardens. That is my profession.
> (ii) to collect dried specimens of interesting plants for study. That is also my profession.
> (iii) to explore unknown mountain ranges, and find out something about their past history, the distribution of their plants, and any other secrets they are willing to reveal. That is my hobby.

In the years that followed, Burma and Assam became very much his own territory. Both countries contained the ingredients which he found irresistible: a rich and varied flora, rugged mountains, fine rivers, dense forest, and a polyglot mixture of tribes. He described them thus:

> By the geographer and naturalist ... Assam and Burma may be regarded as one – a single wedge thrust up between Hindustan and China, blocking the way to Tibet: a wedge

that serves not to unite India and China, but for ever to keep them apart. From the apex of this wedge a range of hills runs southwards, dividing Assam from Burma and thrusting the Lohit River westwards on one side, the Irrawaddy southwards to the other; and from the very earliest times the people who filtered through this back-door from the plateau to the plains were likewise cut into two streams, ever diverging.

It is a measure of the fascination the area exercised over him that he did return to it after the 1926 journey, which was dogged by discomfort, illness and difficulties. When the rains were at their height some of his servants deserted, taking with them the bulk of the seriously depleted stores. This left K-W with no choice but to leave his camp at the Seinghku–Adung confluence and make an eleven-day forced march to Fort Hertz with his remaining servants. It involved a hazardous crossing of the Seinghku River by a rotting cane bridge.

He admitted in a letter to Hinks written at the end of August from Fort Hertz that he would not care to repeat the journey. During the terrible march through the saturated, steamy forest he was under constant attack from myriads of bloodsucking insects. He fell ill, lying awake throughout the nights under the ceaseless torment of the insects. In the dark the forest was made sinister by the phosphorescent glow of fungi, which stealthily gorged itself on the trees. His sickness grew worse when they were wading through the swamps of the Tisang River, where they were plagued by leeches and sandflies.

> ... we had to wade for miles across the flooded valley of the Tisang, so that with the awful heat, and sickness, I could only go slowly, stopping often to rest. It is not nice to be ill in the jungle, with no white man to hold your hand, and to know that you must keep moving at all costs, or fare worse. To rest in a swamp such as the Tisang valley in August is to invite disaster.

The deserters heard of K-W's forced march while they were sheltering in a village. Convinced he was pursuing them, they panicked and robbed the village headman of a hoard of money to pay for their escape. However, the deserters were eventually caught and the money and some of K-W's property were recovered.

After a short stay at the Fort to regain his health and strength, strength, and to gather fresh stores, K-W returned to his camp and the hunt for plants. Despite seemingly endless difficulties, including another desertion and further illness, he made a magnificent collection. Apart from eighty-six rhododendron species, he discovered a superb red poppy, *Meconopsis impedita* var. *rubra*. But the plant that gave him the greatest pleasure was the 'Tea Rose' primula (*Primula agleniana* var. *thearosa*). He called it the 'Tea Rose' primula because its huge pink flowers reminded him of the rose 'Madame Butterfly', a sport from 'Ophelia' (a mutation growing on a plant which produces different coloured flowers to those of the parent plant), which had become instantly popular when it was introduced in 1918.

K-W's love of the primula was due, in part, to the dramatic circumstances of its discovery. He had been forcing his way through dense forest all day before bursting out on to the alpine slopes, where drifts of snow still lay unmelted. Although he was tired he decided not to waste the last half hour of daylight and so made a hurried foray among the alpine flowers. He crossed a snow bridge, and looking up the slope saw a glowing patch of rose pink on the edge of the snow. He was transfixed.

In *Plant Hunting on the Edge of the World* he declared:

> I can recall several flowers which at first sight have knocked the breath out of me, but only two or three which have taken me by storm as did this one. The sudden vision is like a physical blow in the pit of the stomach; one can only gasp and stare. In the face of such unsurpassed loveliness, one is afraid to move, as with bated breath one mutters the single word 'God' – a prayer rather than exclamation. And when at last with fluttering heart one does venture

to step forward, it is on tiptoe, and hat in hand, to wonder and to worship.

And so it was now. I stood there transfixed on the snow-cone, in a honeymoon of bliss, feasting my eyes on a masterpiece.

It was a moment of ecstasy which he never forgot. The huge rose-red flowers clustered into a glowing globe, the only warm thing in that exposed meadow chilled by the patches of still unmelted snow, and loud with the sound of racing water. He also found *Lilium wardii*, which is sometimes known as the 'Pink martagon' or the Tsangpo lily, which he had first discovered in Tibet in 1924.

During this expedition, as in all the previous ones, K-W pushed himself to the limit. He ate indifferently and ignored bouts of fever, with the result that his health became quite seriously undermined, seriously enough to persuade Florinda to travel to the East to meet up with him in Rangoon. He was not pleased with the arrangement. For one thing he intensely disliked being fussed over, but what concerned him most was the cost to the enterprise. The Torquay business had been wound up, and he and Florinda had to depend on her small private income and his precarious earnings from collecting, from his travel books, from articles he wrote for magazines like the *Gardeners' Chronicle* and *Blackwood's*, and from lecturing. What little he had to spare, and it was little enough, had to be invested in new expeditions.

Although financing expeditions was never easy, following the success of the 1924 journey he obtained sound backing for the 1926 and the 1927–8 expeditions from a syndicate of enthusiastic gardeners got together by the British banker, Lionel de Rothschild, who was creating his famous rhododendron garden at Exbury. K-W also received a grant from the government grant committee of the Royal Society, and the trustees of the Percy Sladen Memorial Fund, after which he named the 1927–8 expedition because of the size of its contribution. In addition £600 was put in by his new travelling companion, Hugh Clutterbuck, whose total expenses were about £1,200. Of all his travelling companions, Clutterbuck, the Arctic explorer, whom K-W nicknamed 'Buttercup', became the closest friend. K-W dedicated *Plant Hunting on the Edge of the World* to him with the words: 'My Dear Buttercup, travel books must be dedicated; and to whom could I dedicate this more aptly than to you, who shared the privations, disappointments, and (dare I say?) triumphs of the Assam journey? So you find yourself, for better or worse, godfather to a bantling of sorts. Yours ever, K.W.' When he used the word bantling was he referring to his book as a brat or a bastard?

Their friendship undoubtedly flourished because Clutterbuck was a generous, cheerful, and above all, calm and patient man. He took care not to ruffle K-W even when he was being unreasonable, such as the time he cut a rope bridge when their Mishmi porters went on strike, refusing to carry the loads either forward or back to the village they had just left. It was a foolishly impulsive act for such an experienced man, which enraged the disaffected porters even more, and in the end he had to pay for a new bridge. But throughout the confrontation, which at one stage looked dangerous, K-W was impressed by Clutterbuck's calm.

He was a selfless man, willing to play a minor role without stinting himself. In a letter written to K-W from Thurso in August 1927, he said:

> Of course I am willing to help you to the limit of my extremely limited ability with the survey work or anything else, and will take on as much as possible all staffwork, rations etc., so you can have as much time as possible for your work.
>
> Yes I know the danger of being thrown together on each other's company for months. We must hope for the best, and as you say if the time comes when we get a bit on edge with each other we can always part for a few weeks.

In his early correspondence, Clutterbuck addressed K-W quite formally as 'My dear Kingdon Ward', but before long he was calling him 'My dear K.W.', and the tone of the letters was that of an intimate friend.

Frank Kingdon-Ward

Writing in November 1927 before he sailed to join the expedition, Clutterbuck hoped 'all goes well with you in Act 1 Scene 1 of the new play [K-W was attempting to write a play] – You *are* a lucky man in one thing – you are missing 2 English Christmas's. If I had thought of that sooner I would have accompanied you, even if I had had to share a bunk with a stoker.'

By now K-W's reputation was such that young men were anxious to serve an exploring apprenticeship with him. In September 1927 one of the most successful growers of K-W seeds wrote to him on behalf of a Wisley student called Milligan, the son of a Belfast merchant, who was keen to join him on the 1927–8 expedition. K-W makes no mention of him, so clearly the request came to nothing.

On his way to the East in 1927, K-W paid a flying visit to America. Unfortunately there is no record of what exactly he was doing there, but it can safely be assumed he was there to lecture, and also to consolidate relationships with the New York Botanic Garden, for whom he was eventually to collect seeds, and to make himself known to Suydam Cutting and A.S. Vernay, who used their wealth to support and take part in expeditions.

Clutterbuck joined K-W in India in 1928, whence they travelled to Assam and the northern frontier of Burma. It was not an easy expedition. They had trouble with porters and servants who went on strike for more money or simply deserted, coupled with a shortage of rations. In *Plant Hunting on the Edge of the World* K-W described how they dealt with food shortages:

> ... lunch consisted of a few biscuits, some figs, and raisins, and half a slab of Mexican chocolate. Buttercup had packed these in $\frac{1}{4}$ lb tomato tins, and we standardised this form of lunch ration, which served admirably as long as the figs and raisins lasted. In order to conserve supplies, we enacted a statute that you could only draw the lunch benefit if you were out for three hours or more; but as a matter of fact one was often out for longer than that without drawing it, or at any rate, without eating it. Only on deliberate occasions when we were engaged on a serious alpine climb

did we bother much about lunch; our resources were too strained to make it a habit.

Their rations were to fall so low that they were reduced to scavenging for bamboo shoots. 'We might indeed have eaten voles which abound in the forest, and which are easily trapped; but we were never driven to quite such desperate straits — though I have eaten and enjoyed roast vole,' wrote K-W. They still had supplies of rum, chocolate, biscuits and O.K. sauce: 'excellent things in themselves ... but not a diet on which to climb mountains'.

Despite these tribulations relations between the two men remained unimpaired, even on an occasion which would have tested the patience of the most phlegmatic character. They were camped on a steep hillside, and had gone to a spring at the foot of the hill to fill an old petrol can with water. They took it in turns to haul the heavy load back to the camp. After forcing his way through a rhododendron thicket, K-W, who was carrying the can, set it down to take a breather. He described what followed in *Plant Hunting on the Edge of the World*: '... it slipped and toppled slowly over! For a few seconds I watched it fascinated, but rigid with horror, while trying to regain my breath. Before I could do anything the heavy tin was leaping and rolling down the slope. We caught one flash of the scarlet missile cleaving the air in an immense final bound, and then it became lost to sight and sound.' In fact they did find it, battered but intact, and hauled it safely to camp. At this time they were without servants, and it was Clutterbuck who somehow or other managed to get a daily fire going when the rain poured down.

At the end of the expedition, K-W was able to write of Clutterbuck: 'Staunch, brave, patient, and competent, no man ever had a better companion. Throughout all those months of discomfort, anxiety and stress, no harsh word, no reproach had passed his lips.'

Their friendship continued, with Clutterbuck taking a close interest in K-W's journeys. Some years later, he welcomed K-W back from the 1933 expedition to Assam and Tibet:

> Just a line to welcome you back to 'European Soil'. As soon as you possibly can, we hope to see you in person... Well, well, what a trip you have had. My mouth waters at the thought of all I am going to hear about it all (I hope) when you honour us with a visit – We will not hustle you unduly, as I know you will have much to do these coming weeks, with publishers & Hinks hard on your trail.
> So welcome home Sir and I shall this morning drink your health and our next (and I hope imminent) meeting. Yours ever, Buttercup.

The warmth of the friendship was mutual, never more clearly defined than in K-W's letter following Clutterbuck's death, which he wrote to his widow Elizabeth from the Burma–Indo-China frontier at the end of January 1942:

> My Dear Elizabeth,
> It is with a heavy heart I settle down to write a few lines of sympathy with you at the death of Hugh – my friend Buttercup, one of the two best and bravest men I have ever known. Even in these tremendous times, private griefs for each of us must come, and do come first, and I cannot adequately express my heartfelt sympathy with you and Jasper at this sad hour. I can still see Buttercup marching along, singing, whistling, or talking pleasantly under the most depressing and horribly uncomfortable conditions in the Mishmi Hills; and so I shall always remember him, as a courageous and generous companion who, when everything possible went wrong, never uttered a word of complaint, or gave a sign of distress, lest it upset me further. I hope Jasper is old enough to realise and always remember what a splendid father he had. He can do no better than grow up like him – and may I add, like you. How fares the lad? He must be eight or nine by now! The loss of a father at that age is a sad blow, but with the resilience of youth he will get over it.
> I won't write any more now, but wish you both good luck and years of happiness together.

Kingdon-Ward Territory

The closeness that developed between K-W and Clutterbuck was not repeated in 1930–1 between K-W and his new travelling companion, Lord Cranbrook, an enthusiastic and talented zoologist, who accompanied him to North Burma. His son, the present Lord Cranbrook, recalls: 'My mother remembers there was never at any time a particular closeness between them. Kingdon-Ward did not encourage friendly overtures and my father found it acceptable to make a trip of this sort on such terms. Before they started there was some dispute over the stores, which, as my father feared, did run short.'

The expedition started in something of a holiday atmosphere. They were accompanied to Rangoon and a little beyond by two young women: 'Boo, a dark and handsome Welsh girl with all the spirit of her native hills in her moods, and Betty Rose, in contrast, a modern young English blonde.' Because Cranbrook was travelling with them they were invited to stay at Government House in Rangoon. K-W was not impressed by the British community in the city. In *Plant Hunter's Paradise* he had this to say:

> How correct everything was! Trim bungalows standing back from the road, half-hidden amongst trees, smart motor cars bringing carefully dressed officials (Harrow and Trinity) home from 'office'; other cars taking smart ladies out for a drive; spick and span native ayahs taking the children out for a walk. It was pure Kipling, Simla in little. These government officials do their twenty-one years' service – not necessarily work – and then retire to Cheltenham and visit London once a week, eat curry at the Oriental Club and write letters to the *Morning Post*. That is their life. I preferred my own.

The plan for the expedition was to go to Mytkyina and trace the source of the Tamai River, and from the source to find a way into Tibet and identify the source of the Taron River. But before setting out on that mission there were a few days of picnics on the banks of the Irrawaddy with the girls, where they brewed tea from wild tea bushes, which proved to be disgusting.

On 20 November 1930, Cranbrook and K-W set-off; Betty

and Boo accompanied them, riding in a Chevrolet, for the first forty miles. 'We had been entertained and fêted wherever we went. Photographers had photographed and pressmen interviewed us. We felt grand and significant and enormously pleased with ourselves – until we remembered that we were living entirely on credit. As yet we had done nothing ...'

On the first evening on the trail after the girls had left they were subdued. Their cook, Ba Kai's, first dinner – 'a wet fish he had apparently forgotten to cook' – added to their sense of depression, and K-W felt ill and gloomy. 'The immediate break with friends and home comforts,' he wrote, 'however skilfully led up to, is always depressing. There is no one to rally us now. All our strength must come from within.' While the two men pressed on into the wilderness, Boo and Betty Rose were touring in Burma, cruising on the Irrawaddy, visiting Bhamo before returning to Rangoon via Myitkyina and Mandalay.

In Rangoon unkind gossip was circulating about Cranbrook. In a letter dated 28 December (1931), which she signed, 'your little friend', Boo told K-W: 'Rumours here about Cranbrook are too funny, he is supposed to be a half-wit or a ne'er-do-well whose parents handed him over to you and hoped for the best – also he is engaged to Betty Rose – By the way I have hopes, he sent her a nice lengthy telegram for Xmas – sound him for me – it is £6,000 now and double that later on.' The last reference was to Betty Rose's income. Boo was marriage-broking and one of her main reasons for making the trip to the Far East was to marry off Betty Rose to Cranbrook. The scheme came to nothing.

Although Cranbrook did not find K-W easy to get along with, K-W, on the other hand, was generous in his remarks about the younger man: '... for a good enough cause, he would spring out of bed at any hour of the night you cared to wake him; nor have I met a man more sweet-tempered when roused from sound sleep.' Food, or the lack of it, or the dullness of it, dominated their conversation. Writing to his sister, Winifred, from their camp at Adung on the Burma frontier, he told her:

The rains are now in full swing and we live in an atmosphere of dirt, not squalor. We have enough to eat — but not too much, and a fair variety, though very little fresh food, except a fungus we collect in the forest. Our conversation always comes back sooner or later to the subject of food, our favourite dishes, what we miss most in our diet, and what we will eat when we get back: a sure sign that we don't overeat, and are probably deficient in certain vitamins or virtues. We think hungrily of fish and chips, sausage and mashed, doughnuts, and other simple delights; but this evening we did make ourselves a suet dumpling, and tomorrow night, wrapped in a clean white shroud (my best handkerchief) we are going to boil it alive. Then we cut it in half, and eat it with treacle. We were a bit skeptical [sic] at first because we knew not how much suet to put with how much flour. We only have a pound of suet all told, and not very much flour. However the dumpling when modelled to scale looked cold and clammy as a dumpling should, and as long as it looks suety when cooked, it will be all right.

From the Adung Valley K-W sent ten rupees to a friend, Mrs Thyne, with a request to her to buy Chinese cakes for him and Cranbrook in the bazaar in Myitkyina and send them up in a biscuit tin. 'The Chinese name is *dien shen*, and there are several kinds, a sort of sponge cake, and a rolled hardish cake made of lard and flour (I believe) filled with brown sugar being the best. They are frightfully good, and Cranbrook wants to try them — so do I. We don't get much in the way of sweets, and Barnetts' tinned cakes, although quite good, are very different.' They had just received mail after a wait of ten weeks. It included fresh stores, letters and newspapers. 'Now we have a stock of literature, and something to discuss for the next fortnight. After that we shall talk about food again, and I suppose say all the same things over again.'

It was during this expedition that he first saw the carmine cherry (*Prunus cerasoides rubea*). This was a plant that was to become something of an obsession, largely because it failed to live up to expectation after he introduced it into cultivation. In the wild it was magnificent, with rich carmine flowers, but in English gardens, although the flowers were carmine in bud they opened to reveal a wishy-washy pink flower. When K-W saw it growing wild it blazed with a massed profusion of bloom. In captivity it was sparing with its blossom. He collected it on several different occasions during the rest of his career, but as a garden plant it never matched his descriptions of it in its native habitat. When he first saw it growing close to the Adung Valley camp, he judged it to be the most magnificent hardy flowering tree he had ever seen.

> ... close to our camp, I noticed a big cherry tree about to flower. Two days later it was in full bloom. It was quite leafless and just a mass of blossom, stark crimson. For a minute I stood before it, unable to speak a word, drunk with the glory of it. It was not to be believed. When the everyday world came back to me, I was in doubt for a moment whether I wept, shouted, or said a prayer. Then I turned to Cranbrook and said 'Golly!' in a sort of awed whisper.
>
> The ruby-red flower-buds appear about the middle of March, in compact clusters towards the end of the branches, and the tree is swiftly transformed into a frozen fountain of precious stone. As the buds open the stalks lengthen till the flowers are hanging down. Then the whole tree bursts suddenly into carmine flame. To see the setting sun through its branches when the tree is in full bloom is a thing not easily put out of one's mind.

K-W's finest plants were never out of his mind and his remarkable memory enabled him to remember exactly where a particular plant grew, even after a period of years. A striking example of this was the story of *Cypripedium wardii*, which he told in *Plant Hunter's Paradise*. The locality was the dense forest beyond Fort Hertz.

> The first time I passed this way in November 1922, after crossing the mountains from China by way of the Salween

Kingdon-Ward Territory

Valley and the Taron, I had found a remarkable slipper-orchid in bloom in the hill-jungle. This was three or four days before I reached Fort Hertz. It was a very wet day, pouring with rain. I was not well and had already marched several hundred miles from Likiang in Yunnan. There is no doubt that I was not very observant on that day. And yet I did see the slipper-orchid, a single plant, and I collected it. That evening, when I came to catalogue my specimens, there was no orchid! It had fallen out of my bag, which was crammed with specimens. Regret sharpened my memory, photographed every detail of the incident on my mind. Through four long years I remembered it. I could see the orchid in my mind's eye quite clearly, a glossy chocolate brown slipper, with a white and green standard, and I knew exactly where it grew and what it looked like. As soon as I saw the place, I should halt and say: 'My slipper-orchid grew just there.' Some day I would pass that way again, if only to gather the orchid. And so it was. I travelled this road in April 1926, when I was making for the Tamai River. In August, on my way back to Fort Hertz for a rest, and again in September, on my way through Assam, I passed the very same spot. But I could find no slipper-orchid. Each time it rained heavily, and the undergrowth had grown up breast-high. I could see no sign of the mottled leaves of the lost orchid, and I shrank from crawling into the leech-ridden jungle to search more closely. It is in the winter months that one must search. April is already too late. August and September, in the height of the rainy season, are hopeless.

Another four years passed, and I had not forgotten my slipper-orchid. On December 29th, 1930, I came to the very spot where eight years previously I had found the plant. There indeed was my lost treasure, again one plant! It is a handsome orchid – the stem, bearing its single slipper, about eight inches high. The dorsal petal is ivory white, sharply striped with emerald green, the slipper a glossy coffee-brown, and the narrow lateral petals heavily mottled in ochre and sienna, with bristly margins. This orchid is confined to limestone rocks, which crop out on this range. But now, after finding the first flower, I found dozens. It was the flowering season, and my new *Cypripedium* was as common as the greediest orchid-hunter could wish. I found it on each of the ranges we crossed between the Ti Hka and the Tamai and again farther north in the Tamai Valley itself.

On December 7th, 1931, on the return journey I collected several dozen plants from the original spot, and posted them to England from Fort Hertz. They flowered in 1933, and my Burmese slipper-orchid, which it had taken me eight years to collect, received the name of *Cypripedium* (or *Paphiopedilum*) *wardii*.

Because of Cranbrook's special interest, there was a strong zoological element in the expedition, which gave K-W a great deal of pleasure. Cranbrook guided him in to the seething world of insects. As he wrote in *Plant Hunter's Paradise*:

> We had started a collection of insects, mainly beetles, and this occupation gave us plenty of interest. Whenever we came to a large log of wood or a stone, we gathered round it armed with killing bottles, and proceeded to roll it over if we could. The sudden light startled the underworld into life. Small beetles scurried this way and that. There were vicious-looking centipedes and bloated wood-lice running madly about. Little black scorpions turned up their tails and adopted an agressive attitude. These last were more common inside the wood logs than under them – especially where white termite ants had been at work. Those who are familiar only with English or even with European forests would be surprised at the work performed by the hordes of insects and other creatures which work in the dark in a Burmese jungle. Even at this off-season, prodigious numbers of small sappers and miners were at work. A thick tree trunk is reduced to dust in a few months by a termite army. Nor were all the insects we discovered subterranean.

Butterflies and grasshoppers, dragonflies and beetles were numerous on warm afternoons in every open place.

Ornithology too was an added dimension to plant hunting. 'Whenever I went out plant hunting, I kept one eye on birds and compared notes with Cranbrook in the evening. It was my business to analyse the contents of the crops of Cranbrook's specimens.'

But however fascinating the insects and animals, it was, as always, the plants that drew him and inspired some of his best descriptive prose. Writing of a pyrus, he had this to say:

> At the end of March, in the dense thickets which lined the river bank, pyrus 'Goldbeam' was a beautiful sight. This is a large spread-eagled shrub, with buds of two kinds — inflorescent buds, containing both flowers and leaves, and pure leaf-buds. The former open first. The great black bud is sharpened to a silver point like a polished dagger, as the young leaves come out, closely followed by their attendant flowers. When these leaves are fully expanded, but before the flowers open, the true leaf-buds break, sending out a sheaf of purple plumes. Finally, the flowers open. They are cream-coloured, with amethyst anthers set like jewels on their tiny trembling stalks, and fragrant as meadowsweet. A golden haze of woolly hair envelopes the whole inflorescence, like a spider's web. On the underleaf surface it takes the form of a fine skein, and as it sloughs off in a golden cloud it exposes a layer of silken silver hairs. Eventually these also fall away, leaving only the bald green leaf surface.

At Tahawndam a single specimen of *Rhododendron magnificum* ('Rose Purple') held him spellbound. He later declared that it must have been old 'when King George the first reigned in England ... At a height of five feet from the ground the trunk was a yard in girth. This colossus bore a thousand trusses of flowers, all in bloom at the same time. I have never seen a more magnificent tree.'

But for all their beauty the rhododendrons held a hidden menace. Honey from bees that had been feeding on the great trusses of blooms tended to be toxic. After eating honey brought to them by the Tahawndam villagers, Cranbrook collapsed when he was out shooting, and fell into a river. He was lucky not to have been swept away. One of the servants was also taken ill, as was K-W after eating popcorn soaked in the honey. The symptoms were similar to acute alcohol poisoning. They found that a local mead brewed from the same honey made them extremely drunk.

While K-W could be moved to tears at the sight of particularly beautiful flowers, or spend hours lost in wonder at a view of snow-clad mountains, or the swirling jade-green waters of a Himalayan river, his habitually rather grim, tight-lipped expression could be somewhat intimidating. It was as though he was deliberately holding himself at arm's length, an impression that Cranbrook took away with him after thirteen months of travel with K-W. It created a somewhat uncomfortable atmosphere.

It was an atmosphere that the 23-year-old Ronald Kaulback felt keenly at the beginning of the 1933 expedition, during which he accompanied K-W to Assam and the border with Tibet. (The third member of the team, R. B. Brooks Carrington, took a more minor role.) Kaulback found their first meeting in London quite unnerving, and their train journey from Calcutta to Assam was, he recalled: 'A little bit uncomfortable. He was not a comfortable man to be with until you knew him.'

Kaulback always had the feeling that K-W had his eye on him: 'I always felt he was looking at you in a critical way from the outside, no matter what was happening. I think he was partly examining your intellect, but mostly to see if you were capable of doing the job.' Kaulback's job was surveying, map making, and commissariat duties. He collected reptiles as a private interest.

Kaulback did not recall K-W ever losing his temper over mistakes. 'He just made you feel a bloody idiot, which you were.' Like Cawdor, Kaulback found K-W's withdrawn manner irksome:

> He was, to start with at any rate, until I got to know him, a very, very difficult man to travel with, simply because he could so easily fall into total silence, which would last for two or three days. Not a damn thing. He might say 'Good Morning', otherwise he would just march along,

and at the end of the day sit down and have a meal, but nothing, not a word.

It would be a bit awkward. If you tried to make conversation you would be met by a wall of silence, or a curt reply. After I had been with him some weeks I realised it was all right, and it didn't worry me any more, but it worried me immensely to start with. After a few days he would snap out of it and be great fun.

Kaulback believed this apparent moodiness was the result of worry, of constant planning ahead in his anxiety to make a success of the expedition. For a plant hunter who wants backing for future journeys there is no room for failure. He cannot return empty-handed to his sponsors, and if he does he must return their investment.

When he was not distracted by worries or the demands of his job, K-W would go to great trouble to explain to his companions the characteristics of the country they were travelling through, and in the evenings, after dinner, he would initiate almost childish parlour games. Kaulback recalled: 'He would suddenly say something like: "let's think of all the places, one after the other, beginning with P and the first fellow to peg out has to buy a drink when we get home".'

To start with, in his job as quartermaster, Kaulback found the Mishmi porters very difficult to handle — 'bloody-minded blisters' he called them, and he developed a huge admiration for the way K-W dealt with them. At one point when the porters dumped down their loads and refused to go on without more pay, it was K-W who took over. Kaulback recalled:

> He had a tremendous and magnificent scarlet sweater, made for somebody six times his size, it was huge, it came down to his knees; bright scarlet. He realised early on, I think, that that sweater was going to be of great value to him, because the Government used to issue well-behaved headmen of the Mishmi villages with scarlet flannel jackets, and this enormous sweater was obviously far better, far more important, far more valuable than a flannel jacket, so

he always used to put it on when he was talking to Mishmis.

He never got obviously upset with the locals, and I think that calmness he had was of much value. If there was trouble he would appear looking a little bored, and sit down and just wait for them, and look so uninterested it would knock all the fire out of them. He was a very strong-minded person. Once he had decided this was the way to do a thing, then that was the way he did it, no matter what.

Later when he and K-W had parted on the Tibetan border, Kaulback was travelling through Burma when he was faced with a porters' strike. Again, they wanted more money to carry on. 'I thought, "What would Kingdon-Ward have done?" He would have said "Don't get upset, don't pay any attention to these perishers, call their bluff." So I said to the porters: "You don't get paid now. The agreement is that you get paid at the end of the journey, but you can go back now if you want to, with my blessing." They went off shouting and yelling, and then carried on. That was Kingdon-Ward's way.'

From time to time Kaulback experienced what seemed to be a coldness and lack of sensitivity in K-W. On one occasion he had been sent to look at a remote monastery about fifteen miles from where they were camping. He was kindly received by the monks, who gave him a meal and tried to persuade him to spend the night, but afraid that K-W might worry about him he insisted on returning and was lent a pony for the journey. The pony simply wandered round in circles and he ended up leading the animal, and falling into several streams on the way, before reaching the camp drenched and exhausted. He woke K-W to say he was back. The only reaction was a testy: 'Why did you wake me up? I thought you would be spending the night at the monastery.'

In Kaulback's opinion, K-W was a shy man who was more at ease with his own company. 'If he was utterly alone, with nothing but coolies and a couple of servants to do his cooking and make his bed, that was his real happiness. He used to take on

Frank sitting on his mother's knee

Professor Harry Marshall Ward,
Frank Kingdon-Ward's father

Frank with his sister Winifred

Frank (*left*) with Kenneth Ward on their bicycles

Winifred in her late teens

Frank, photographed soon after he arrived in Shanghai

In Mesopotamia, 1917

The young explorer

Rhododendron nuttallii, Adung Valley

(*Opposite page*) K-W seeing his porters safely over the Ata Chu

Setting off on the plant trail

Tibetan minstrels

Primula agleniana

Lilium nepalense, growing at 6,000 feet in the Assam Himalaya

Orange-flowered dendrobrium, the largest orchid in North Burma

Lilium wallachianum, Assam Himalaya

'White Vanda', an epiphytic orchid found in North Burma

Collecting in North Burma

K-W with a dendrobrium orchid

(*Left*) *Michelia doltsopa* and temperate rain forest in the Adung Valley

(*Right*) With Jean outside their camp

(*Inset*) *Lilium mackliniae*, named after Jean

Jean pressing plant specimens

Base camp in North Burma

Mishmi porters carry Jean to safety after the earthquake

"Popperfoto"

F. Kingdon-Ward

somebody like me, partly for financial reasons, and partly to do the jobs he didn't want to do himself, such as map making. He did all the botany. There is no question he was a loner, he liked to be alone.'

K-W's impatience with what he saw as incompetence would rub people up the wrong way, even though his sharpness contained no deliberate unkindness. The third member of this expedition, the young R. B. Brooks Carrington, had joined in order to make a colour film of their progress. Sadly that project failed. Brooks Carrington's operation became a source of irritation. When he wanted to stay in a location until the light was right, K-W wanted to know why he hadn't shot the film the day before, and accused him of holding up the entire expedition.

Despite his apparent coolness, K-W did have an enormous sense of fun, remembered by his friends in England during his spells on leave, and for ever enshrined in a bizarre event during the 1933 expedition. It was Easter and he and Kaulback were invited to dinner by the Governor of Zayul. K-W decided they should do a party turn by way of thanks and took his ukelele along. After a magnificent meal and lashings of powerful Chinese spirit, he launched into a medley of minstrel show tunes like 'Swanee River', 'Old Black Joe', and 'Ukelele Lady', while Kaulback danced the Black Bottom and the Charleston.

'It was one of the only times I saw him laugh; he was laughing away like mad at the table, plied with that god-awful Chinese spirit, and twanging away at this evilly played ukelele. He was singing and I was dancing. We were the success of the evening, especially with the villagers; the villagers were thrilled,' Kaulback recalled.

K-W remembered the evening in *A Plant Hunter in Tibet*: 'As a farewell gesture I played the ukelele to our delighted host, while Ronald Kaulback performed an extempore dance which would have gone down well enough in a Piccadilly night club, but which created a huge sensation in official circles in Rima. The governor was not less delighted than was Herod with Salome...'

While Kaulback grew to admire K-W enormously, he was never able to get on to the same terms of friendship as had developed between K-W and Clutterbuck. But after more than fifty years Kaulback's memory of him remained sharp:

> He was always terribly fine drawn. You always looked at him and thought, 'I would like to give him a really good meal.' He looked as if were you to land him a clout he would fall flat on his back. But he wouldn't. He was the toughest fellow I ever knew. We would have a relatively tough march, it might not be far in miles, but it was in hours. We would get into camp in the evening, and if the weather was fine he would look up the mountain with his binoculars and see something, and without another word would be off for two hours climbing to see what it was, and come down again, while we were happy to sit down.
>
> But there would be other times when he showed no enthusiasm. He would be flat. He always felt he could have done better, and should have done better. He was never, never satisfied. At the same time I do think he felt he was not properly recognised; that people should have flocked to support his trips, instead of just meagerly saying we will pay you for the seeds.
>
> He always used to tell me that I was an old-fashioned kind of explorer, but he really was in the mould of the old-time mid-nineteenth century explorers. Although he was not a great geographer, at least not as great as he was a botanist and plant collector, he went to a lot of places where nobody had been before.
>
> He was tough, really tough, and a great man.

7

From the Thames to the High Himalaya

IN 1934 K-W was in England on one of his rare home leaves. During his fourteen years of marriage to Florinda, he was with her for little more than four of them. It is doubtful if they ever really got to know one another. With long separations their lives drifted apart almost before they could be harmonised.

To try to understand why their marriage was doomed it is necessary to go back to their wedding in 1923. As we have seen, the proposal, by letter, came from Florinda, and was presumably accepted by letter from K-W, although no such letter has survived. The wedding itself was a swift affair in April 1923 at Kensington Register Office. They only had a few months together before he was off early in 1924 to the Tsangpo.

There is no doubt that Florinda worked hard at the marriage. In particular she went to great efforts to provide him with a superbly run home to return to, and that is probably one reason why the marriage lasted as long as it did. When K-W was on leave he took a keen interest in his home, particularly the garden.

At Hatton Gore, the Harlington house, he constructed a rock garden, designing it to look like a bend in a river ravine in the Himalayas. It was built from York stone from the demolition of the old Bank of England building, acquired by a young official of the bank who, with his wife, rented a flat in the Kingdon-Ward house. The garden was planted with species raised from K-W's collected seed.

Frank Kingdon-Ward

Growing plants from his own seed was one of the excitements of the household, and everyone was called out to see the newly germinated plants. In all this he was greatly helped by Florinda, a keen and skilful gardener, and the two gardeners she employed.

With the birth of their two daughters, Pleione on 21 March 1926 and Martha on 11 March 1928, they moved from Hatton Gore to a larger house, Cleeve Court, at Streetly-on-Thames. It was to be the last home they would share together. It was a large Edwardian building which had belonged to Lord Craigavon, the Prime Minister of Northern Ireland. The bedroom verandas looked out over the river, and it stood in seven acres, with a lake and walled kitchen garden. It was the kind of grand setting in which Florinda flourished. Her staff included a butler and cook, and a nursemaid for the children;

There is no doubt that Florinda found the money for the house, but quite how is something of a mystery. She certainly had a small income of her own, and a family trust which she managed to draw upon. She ran an advertising agency, and with her magnificent hair modelled for a shampoo advertisement. She dabbled in buying and selling property, and tried to undercut the coach companies running services into London by setting up a network of private car owners who would motor people to London for a small payment.

There was also the income from the young couple who rented a flat at Hatton Gore; they came in on the Cleeve Court purchase by buying the lodge gate cottage. Another flat at Hatton Gore was let to a very close friend of Florinda's, Eve Hadfield. She had run a successful PNEU (Parents' National Education Union) school at Maidenhead, and she also moved to Cleeve Court, probably contributing towards the cost of buying it.

On top of running the house on quite a substantial scale, there was the cost of the entertaining laid on when K-W was at home.

His own lack of earning power, and Florinda's apparent extravagance, caused K-W constant worry. Apart from trying to support a wife and two children, he was also making an allowance to his sister, Winifred. His letters apologising for delays in sending her money indicate how difficult he found it to keep up the instalments. In 1932 he wrote to her to ask for a cheque for £32 which had been overpaid to her by his bank, because 'I am rather badly overdrawn again'. She sent the money, and he wrote:

> Many thanks for the cheque. I am sorry I have had to ask for it, and would not have done so, merely treating it as an advance of next quarter, only poor Florinda's affairs were not very satisfactory, and I had to draw a sum which I don't think my bank would have stood for at the moment.
>
> All hope of not overdrawing my current account again has already been abandoned, but the rise in securities has to some extent compensated that. Still, my bank won't allow me much of an overdraft now, and it *won't* help you any if I do go bankrupt.

The young Bank of England official who bought the lodge gate cottage at Cleeve Court, and got to know K-W well, believed 'that one of the things that hurt Frank was that Florinda seemed to spend money senselessly, whereas he led such an austere life. She had no understanding of what he would have considered frippery.' When Florinda was away K-W would often call on his young neighbours at breakfast time. 'All he ever wanted was a bowl of bread and milk, which he would eat while walking round the kitchen.' His daughters Pleione and Martha recall breakfast times as being rather glum occasions, dominated by K-W's grim silence, although Martha does remember bedtimes when he would sit on her bed and sing to the playing of his ukelele.

Quite often he would call in at the lodge gate cottage and haul his young neighbour off on a long walk on the nearby downs. 'They were quite chatty walks, and we would talk about almost anything, but never go deeply into anything. I was always conscious of the fact that he had a much better mind than I had,' the neighbour said. On these walks K-W would make rather ponderous botanical teases:

> I remember walking with him one lovely summer evening, and there was one of those little toadsflax you get on

chalky downs; we were walking and talking, and Frank suddenly saw a patch of these which were as common as dirt, and he said: 'My God, what's that?', and I said: 'Come on, Frank, don't be silly, it's a linaria', and he replied: 'I've never seen anything like that. If I'd seen that out in the Himalayas I would have brought it home!'

It was a leg pull he often played. A friend, Constance Levinson, remembered sitting next to him on Beachy Head, near Eastbourne in Sussex, when he suddenly picked a daisy and declared: '*Bellis perennis*, now that *is* rare!'

He seemed far more relaxed with his friends than with his immediate family, with the exception of his sister, Winifred. At country weekend houseparties he was remembered as being at the centre of all the fun, flirting with the single girls, and joining enthusiastically in the somewhat childish games like Murder, which were popular in the 1930s, and thoroughly enjoying any adulation that came his way. He was not averse to showing off occasionally. A fellow guest remembered coming down in the morning to the large hall of the house where they were staying to find the floor covered with maps, and K-W on his hands and knees plotting the course of some future expedition.

At home the atmosphere between Florinda and himself was quite different. His neighbour recalled: 'Florinda always looked after his creature comforts, although they were never demonstrably affectionate to one another. There was what one might call a civilised, polite, but not affectionate, relationship. Florinda was too high-powered. I think he would have preferred a cosier wife.'

The stilted relationship between K-W and Florinda, the sudden chills that would descend on the household, the long periods he would spend locked away in his study ostensibly writing or planning the next expedition, were simply the symptoms of a collapsing marriage. During his 1934 leave they came close to a divorce. Exactly who wanted it is not clear. In a letter to Winifred in 1937, he wrote: 'The matter [divorce] was discussed three years ago, in a certain amount of heat, and shelved on my part because I thought the children were too young.'

From the Thames to the High Himalaya

In fact it was easy enough to put off the inevitable, because in February 1935 he left for an expedition which was to take him among the headhunters of the Naga Hills, and eventually across southern Tibet. He had been engaged for his experience and knowledge of the area by two wealthy Americans, both leading naturalists, Suydam Cutting and A.S. Vernay. Suydam Cutting was a New York millionaire, and the British-born Arthur Vernay had made his home and career in America. Also part of the expedition was J.K. Stanford, a Burma civil servant and a distinguished ornithologist. The zoological and ornithological expedition was organised to collect for the American Museum of Natural History, and produced thousands of specimens.

The first part of the journey was by raft on the Chindwin river, and was from K-W's description quite idyllic. His diary entry of 23 March 1935 recorded:

> We drifted on lazily. Blue skies, bluer hills in the dim distance, green water, backed by the green of jungle, yellow sand, and here and there a red sandstone cliff. Porpoises played alongside in the river.
>
> Halted towards the evening for tea, and cooked supper. Then on into the warm dusk. The stars came out slowly. After I fell asleep, and when I awoke the moon had risen. Lights on the other rafts drifted by. Wisps of cloud writhing over the river. Big rafts of cane going down the river with houses on them.

And he added wryly: 'Where is the market for all this cane? Our public schools perhaps!'

In the evenings Cutting and Vernay yarned about former journeys with tales of leopards stalking into bungalows in Assam and lying under beds; of bat caves in the Nergui archipelago; of gorillas in the Cameroons; of deadly mangrove swamps in Borneo.

From the pleasures of the raft they transferred to a ramshackle wooden-bodied Chevrolet bus with a leaking radiator, tyres worn down to the canvas, sidelights that did not work, and a single faulty headlamp with a broken glass. It took them three and a half hours to cover eighteen miles into Imphal. Finally they made a triumphant entry into the town with a man perched on the wing holding the lamp into its socket to keep a shaky contact together. Nothing could conceal K-W's joy in being back in his kind of country. He wrote in his diary: 'Sometimes I have almost wept for joy at the sight of the jungle, green and serene, the softly undulating water, smooth hills flaring out to the distant plains. Now I greet the snow-covered hills and the dark coniferous forest with the same joy.'

Of course the expedition was not free from trouble. K-W was particularly plagued by fleas, not helped by the fact that his Keating's powder had been ruined by the damp. He was forced to powder himself with his plant and insect preservative, which burnt his skin: 'However, it doesn't preserve the fleas. They conk out quite quickly.' But there were compensations as he also described in his diary: 'To come back after a six-hour climb with plenty of material, swill oneself over with hot water, change into dry clothes, and sit down to a well-earned tea gives one a feeling of great satisfaction.' And there were the exceptional plants like the Himalayan forget-me-not, which he described in *Assam Adventure*:

> ... *Chionocharis hookeri*, the Himalayan forget-me-not, is one of the most heavenly of the Himalayan plants. It bulged out of the hard hungry earth, glistening silver-grey, each rug-headed cushion encrusted with large turquoise-blue jewels, shimmering like stars in the lilac dusk. Imagine a dome of coral, with a turquoise set in every pore, the whole mass forming a mosaic of blue and silver amongst the pewter-grey micaceous stones. That will give you some idea of *C. hookeri*. These cushions grow very slowly. I think many of them must have been a century old.

There was the exciting moment when he reached the great range of snow-covered mountains that he had first glimpsed from the Tsangpo country during the 1924–5 expedition, but had been unable to reach because of bad weather. Now he found himself in a valley looking up at this staggering range of mountains. Five glaciers hung on the jagged slopes. Snow peaks, which he estimated rose twenty thousand feet into the sky, were like the spires of some gigantic cathedral.

> If any lingering doubt remained, it was now finally set at rest. We were in the heart of the mysterious snowy range of Pome. The lost range was found! The enormous flight of bergs which Cawdor and I had seen far off, eleven years earlier, arching across the world for over a hundred miles, a glittering skyway joining east and west, was here in front of me. No wonder I felt uplifted!

He was transfixed by the beauty of these Tibetan mountains, which were now being stained with violet as the sun set.

> ... everything in that light was ethereal, almost spiritualized; and presently when I heard the sound of distant song I was not very surprised.
>
> There was a solemn hush about the mountains with their changing lights, the ghostly peaks shining through the oncoming darkness, the red shafts from the setting sun which still sprayed through the western wall and caught the tops of the spires opposite, and the abyss of the Tongyuk River far below. I was in that fanciful mood, when one might hear voices singing. Perhaps I was half asleep; certainly it was all like a dream, if a dream come true.

In fact the singing came from a band of pilgrims trying to frighten away evil spirits in the gathering darkness. Accompanied by his two Darjeeling servants, Tashi Thondup and Pemba, and local porters, K-W left Cutting and Vernay in Assam to trek eight hundred miles across Southern Tibet through largely unexplored country — an area which was botanically virtually unknown. He became the object of considerable official suspicion on this journey. He was in trouble for travelling without the correct passport. One man in particular, a Colonel Yuri, believed he was a Bolshevik *agent provocateur*. Ironically K-W was mistaken on one occasion by a group of Pombas for Colonel Yuri himself. They sat him down on beautiful carpets by the Po Yigrong, served him buttered tea in a silver-lined bowl, and plied him with fruit and *tsamba*. He then played a small joke on a group of soldiers on their way to join Colonel Yuri by holding a rifle inspection and ticking them off for the filthy condition of their Lee Enfield and Manlicher weapons. He was also mistaken for a bandit called Wush Mapu, who was wanted by Lhasa in connection with the mysterious and amazing disappearance in Tibet of three thousand men.

Geographically the journey was a huge success — he surveyed and mapped unknown territory — and botanically it yielded a rich harvest. K-W's description of just one hillside in *Assam Adventure* gives a clear idea of what he discovered. '... the slope was moist with boggy hollows, and spangled with millions of flowers of all colours. There were solid golden carpets of *Caltha palustris*, drifts of white or violet anemone, crimson, yellow and pink pedicularis, *Nomocharis nana* and crowded colonies of sulphur-yellow *Primula alpicola*.' In some shallow pools he collected aquatic plants which, he said, 'have an almost worldwide distribution. I collected water buttercup (*Ranunculus aquaticus*), juncus, spartium, *Hydrocharis morsus-ranae*, hippuris (mare's tail), and a grass.'

The rhododendrons, too, were magnificent and their colour exploded throughout the superb scenery, as in his description in *Assam Adventure* of his first glimpse of the sacred lake, Tsoga:

> Just as I reached a grassy alp surrounded by seas of rhododendron in strident bloom, the veil of the cloud was rent, and in a flash there was revealed, a thousand feet below, one of the most beautiful and inspiring sights imaginable. More and more bright grew the scene, like a swift flame, colour and form harmoniously blended. Not that it was tame in peaceful decay, by any means; it was grand without being savage. I was looking straight down on to Tsogar, sacred turquoise lake. The fretted cliffs ran rivers of fretted ice; the largest glacier reached the edge of the water. On its near side the lake was girdled by an emerald-green arc of rich pasture, dark spotted with yak. There was a smaller lake in front of the main lake, and the stream flowed from one to the other and so away down a

> valley to the south, to join the Tsari River. In the foreground billows of rose-purple rhododendron blossom tossed on the breeze.
>
> There is nothing very remarkable, still less mysterious, about Tsogar. It is just an ordinary glacier lake, half silted up; you may see plenty such in Switzerland. Yet this far corner of Asia exercised a fascination over me; I felt spellbound. Well I knew that no white man had ever set eyes on the sacred lake before.

The expedition took a tremendous physical toll on him. He was nearly fifty, and with the weakness which followed every bout of fever, he began to wonder whether or not he was reaching the end of his travelling career. In a letter written to Winifred from Assam on 14 April 1935, he described what he had to endure.

> I am having an attack of malignant tertian malaria. Today is a 'free' day, but two attacks within three days pulls one down, and though I am not in bed, I have to take it easily, and naturally I haven't much of an appetite. However the worst is probably over; a third attack is due tonight, and if it is quite mild, I shall soon be cured. Malignant tertian is easily and quickly combated, but it wrecks you for the time being – a week or so.

Another letter to Winifred – he occasionally calls her by a childhood pet name, Vinny – clearly describes the toughness of the trek across Southern Tibet:

> ... as it is now five months since I had any letters or news, I am quite in the dark about you, and will have to tell you a few things about myself. I have kept fairly well, all things considered. I was down with fever twice after starting on this trip, but since May have not been unwell enough to take to my bed. When I look back and consider that in five months I have travelled over a thousand miles [he seems to have added an extra two hundred miles] in Tibet, much of it through unexplored country, and crossed over 20 passes, 15,000 to 17,000 feet high, and that I am in my

50th year, I don't think I need complain! Up to the middle of August I did pretty well. After that, the rather hard and certainly monotonous travelling began to tell on me. It doesn't sound particularly dreadful to be without tea, jam, sugar, biscuits etc., for a bit. But it is surprising what a difference it makes to have even a small ration of these little luxuries every day. Well I struggled over the grim passes into Gyanda (a village about 120 miles from Lhasa on the China road) expecting to buy empire products. However Gyanda has gone down in the world instead of up. There was one shop in 1924; now there are none. I had only to cross one great range to reach the Tsangpo – another landmark; but it took eight days, long marches too, and when I reached the Tsangpo I was so done I had to rest two days.

Eventually I got back to my base north of the Himalaya, with all my booty and discoveries. I had been away just 90 days, 68 marching days, and 22 halts – not exactly rests either. It was now the end of September, and I had to start for India, having promised to be back in Assam at the end of October. It was necessary to be going anyhow, as I have no gloves or mitts or indeed any real winter clothing. We had gorgeous weather, but of course the wind was shattering – you can't imagine what it is like at 17,000 feet until you have experienced it. What price the ridge of Everest at anything over 23,000 feet! or on the polar plateau at 10,000 feet! I'm glad I'm not one of those heroic birds. We crossed the first pass so late in the afternoon that we weren't able to descend much below 16,000 feet. I did not get much sleep that night, the temperature inside my tent being 16 degrees below freezing point. I was as rigid as a corpse – I very nearly *was* a corpse I think! However we are now down in the forest, having said farewell to the plateau. Only 14 marches from the plains with eight ranges to cross, nothing very terrible, two or three passes of about 15,000 feet, the rest 10,000 to 12,000 feet. I can do *that* all right. Fourteen marches sounds quite a good way; but at the furthest point I reached in Tibet I was 50 long marches from the plains of India — six or seven hundred miles. Besides I shall be out of Tibet in a week now, and in the frontier tract. When I reach the plains, I am only 30 miles from our outpost — it will be lovely to walk on the flat again, still more lovely to sit back in a chair, and read letters and newspapers.

If I can get news of my arrival through to the outpost, they will send a car to meet me, and I shan't have to walk that 30 miles. But I don't much care; I can do it in a day, or stroll and take two days. Anyhow I have been successful in my quest on the whole, and have got some good plants, as well as other things. I explored an unknown river to its source for a hundred miles through fabulous gorges [the Tongkyuk] till I came to incredible high peaks and glaciers. Great fun it was, but exhausting.

For all that he looked forward to returning to civilisation and the company of fellow Europeans, he found it difficult to adjust from the ways of solitude and the lack of any real conversation. In a diary entry for 30 October 1935 he records: 'First taste of social life. Words do not come easily, nor do I want to play tennis. It is nice to have hot baths and clean clothes again.'

Despite the discomforts, illness and many dangers that he had to tolerate in the wilderness, there is little doubt that K-W felt more secure, more in command, in such places. He writes of the 1935 expedition with a kind of ecstasy, as though out of gratitude for those unexplored mountain ranges and deep gorges that so effectively protected him from a marriage and family life, both of which were disintegrating. Whether consciously or unconsciously, he was creating an unbridgeable gap between himself and his wife and daughters.

A letter he wrote to Pleione and Martha from Assam in November 1935 could just as well have been a hasty note written by a father to his daughters while away on a short local business trip. Certainly it was not one to be expected from a father who had been separated from his children for many months, and it was quite unlike the long, rich letters he wrote to his sister, Winifred. In five brief paragraphs he thanks the girls for their 'charming letters', congratulates them on some drawings they had made for him, says he has been on a river steamer on the Brahmaputra, and ends: 'Now I must go on with my work, as it must be finished by a certain time. Much love and kisses darlings from your affectionate daddy.'

In 1936 he returned home, but any hope that the separation had healed the breach between K-W and Florinda was soon dispelled. Even the most determined optimist would have had to admit that there was nothing left of their marriage, not even the pretence. Something of his attitude may be gleaned from the following letter, written eight years after his own wedding in reply to a young woman friend who had written to invite him to hers:

> So you want me to come to your wedding! I really ought to say thank you for so kindly inviting me, but having spent 20 years carefully dodging weddings (including my own), I am not so keen to come as you might suppose. The fact is I think all that ecclesiastical ritual rather bilge; and the modern wedding is as gloomy as a funeral.
>
> My place is really in the strangers' gallery outside the church, where I can stand incog. dressed in any old Moss Bros. reach-me-downs, and heave rice (uncooked for choice). All this time you will think I have forgotten to congratulate you; but that is only because I feel it is he who ought really to be congratulated. A fatuous custom enjoins us to distribute congratulations like autumn leaves in Valhalla – or was it Vallombrosa? I forget.

When the breakdown of the Kingdon-Ward marriage became irretrievable in 1936, Florinda moved her household to an address in Regent's Square, while he stayed at Cleeve Court. He did not remain there for very long, and in February 1937 he was heading East once again. He wrote to Winifred from Aden where the SS *Castalia* put into port. It was a sad, slightly querulous letter, in which he refers to an apparent disagreement he had had with his sister.

From the Thames to the High Himalaya

> My Dear Winifred,
> It is some time since I heard from you, but for my part I prefer that you should write at longish intervals when you feel you have something to say rather than more frequently, from a mistaken sense of duty, when you haven't.
>
> But I have heard of you, from Eve [Hadfield], who read out a 'long long trail awinding', of mildly abusive clichés — 'hot under the collar' was one of them. They did not even accurately describe my feelings. Anyhow personally I prefer good round abuse with a literary flavour, so long as it is original. I should enjoy being called, for example, a 'bottle-nosed turbot who had worked a fin loose', or a 'smudge-eyed herring' — or anything slightly new even if untrue.

And then he came to the real point of the letter, and it reveals much about his marriage:

> Well, I am writing now because I have some news, namely that I am on my way to China — unless stopped by unforeseen complications in Burma. Anyway I *am* on my way to China. How long I shall remain there I can't say, certainly a year, possibly two, as I am not in any particular hurry to return to England — at least not in my present mood. Which reminds me that I have another piece of news — it may not be news now — namely that Florinda and I are being divorced. That bold statement may give a wrong impression. More accurately *I* am being divorced. Even that perhaps hardly gives the idea. The matter was discussed three years ago, in a certain amount of heat, and shelved on my part because I thought the children were too young ...
>
> From this you will infer — quite rightly — that it is a connived divorce, not, in law, a collusive divorce, but equally illegal!
>
> This year the question again arose, in a rather calmer atmosphere, and was approved by both parties. So King-

don-Ward v. Kingdon-Ward became a fact. The suit was, is, undefended, and in due course a *decree nisi* with custody of the children will be pronounced in Florinda's favour. No sooner was the matter in the hands of the lawyers than Florinda and I became quite good, even warm friends again; which is all we ought ever to have been perhaps. If you ask me *why* I am being divorced, I'm afraid I can only give the jejeune reply. Because I was unhappily married. Also I suspect Florinda was. I think it very probable – I should say I think it highly unlikely that either of the parties will venture again into the perilous and uncharted seas of matrimony. For my part, I am too old, too wedded to exploration, too poor, and perhaps too wise. Naturally I have to continue to support or rather contribute to the support of my wife and children.

But it is not altogether unpleasant to feel *free*. Florinda could marry again of course easily enough. She has a host of admirers, she is still young. But she says, and I believe her, that she is not in love with anyone else, and that she does not wish to marry again. My own impression is she is not the marrying sort – too strong-minded. At present she talks as though once the divorce is through, all will be as it was, and that Cleeve Court will be my home, and so on and so forth. I don't know. To me a divorce is a divorce. We shall always be very good friends, and of course I shall sometimes go to Cleeve Court, and Pleione and Martha are always my babies. But I doubt if there is more to it than that. Nothing would induce me to live at Cleeve again, since the fact that I was requested to was no small part of the reason why I asked for a divorce. What a life, what a world!
Love from Frank.

Unlike K-W, Florinda did not indeed remarry. Instead she devoted her life to her daughters, to business, and at one stage to politics.

The grounds for the divorce were adultery on K-W's part, and, indeed, the petition claimed there had been a long-running affair with a woman in Devonshire. But judging by his letter to Winifred, it was a purely contrived adultery. In those days it was quite common for a co-respondent to agree to spend a night in an hotel room, and give the evidence in order to make a divorce quick and painless.

In 1937, five days before Christmas, and after months of arduous travel, K-W was depressed and unhappy. It was a mood that often overtook him at this time of the year. He wrote in his diary:

> Mail arrived – no English letters for me. Florinda, who at least used to write to me in her spare time (if any) has now it seems decided to write to me (if at all) about once a quarter. She never did take any trouble to write to me for special occasions like Christmas – far too busy with the world's work, or buying 'real' houses (though she now lives in a shabby villa) – or send me a Sunday paper after the boat race, or an international rugger match. But why worry!

Having decided at the beginning of 1937 that his marriage was finally over, and that he would never again set foot in Cleeve Court, on Christmas Day he had a complete change of heart. He wrote in his diary: 'Last night I made up my mind to quash the divorce (absolute) if possible on grounds of collusion. Reasons – free access to my children, and a *right* to my study at Cleeve Court. God knows what will happen if F gets a free hand with the law on her side.' But by New Year's Eve he had argued himself into a calmer frame of mind; an acceptance of what he knew to be inevitable. His diary entry for that day reads:

> Although happiness comes from within and not from without it is impossible that environment, friends, enough to eat, and so forth, should not have *something* to do with it.
>
> Situated as I am living a healthy life, in these beautiful and every-varying surroundings, feeling well and strong, with enough to eat, and no very awful tragedy hanging

> over my head, it is really *ridiculous* for me to feel so unhappy, ill-used, and sorry for myself! It is quite artificial – a wrong mental outlook. I could be perfectly happy if I wanted to, firmly make up my mind to do my duty, and let everything else take care of itself. *I am* making an effort to curb my ill-humour for a start. I believe that anyone who had faith in the continuity of existence, or reads a book like, say, Upton Sinclair's *Jungle*, or reads the newspaper reports of Spain, or China, or the flooded States of the USA, or any similar human calamity, *must* give thanks that he is spared such horrors and rid himself of the cancer of discontent. Suppose I were to die in one of my fits of terrible depression (or worse still, destroy myself) after a period of bitter discontentment because I am not famous, or do not make money, or am not the centre of a glittering throng holding high office! And I believe I *have* been discontented and rebellious for all these absurd reasons, and others equally absurd. I might wake up to find myself in the stockyards of Chicago in winter or unemployed and starving, or in a muddy and bloody trench in China!
>
> So if I make any good resolution for the coming year it must be to keep on trying. To get a firmer grip, on myself. To be less obstinate and pig-headed. To be more resolute as to the difference between right and wrong. To try and understand and sympathise with others. So the year ends.

And so also did his marriage, bitterly. When he left Cleeve Court early in 1937 he left it for good, and, indeed, was never again to have a home of his own. In January 1938 he wrote to Winifred:

> I don't think there *is* anything much to be said now! I have heard very few details, and only heard of the successful result of Florinda's petition from friends in Burma! Lately I have heard that I was splash-headlined in the gutter press – Famous Explorer Divorced etc. – for a collusive divorce discussed years ago, and finally gone through

mainly in Florinda's interest (since she preferred divorce to separating, and divorce to yielding an inch to my few reasonable requests), it's hard luck that I should be pilloried in the press as the villain of the piece.

K-W looked back on 1937 as an exceptionally difficult year, not only in his private life, but also in his professional activities. He had raised the backing for a plant-hunting and collecting trip to Yunnan in China. It was either a foolhardy project, or one conceived out of naive optimism, for China was in the grip of cataclysmic events which were shaping the entire future of the country. Sun Yat-sen was dead, and with him the intellectual idealism of a liberal democracy for the former empire. The real struggle now was between the Communists, receiving considerable support from Russia, and the Nationalists.

Stalin had attempted to promote a coalition between the Chinese Communist Party and the Kuomintang. He declared: 'Such a dual party is necessary and expedient, provided it does not restrict the freedom of the Communist Party to conduct agitation and propaganda work, providing it does not hinder the rallying of the proletarians around the Communist Party, and provided it facilitates the actual leadership of the revolutionary movement by the Communist Party.' Naturally such demands were totally unacceptable to Chiang Kai-shek, the leader of the Kuomintang, himself more of a warlord on the grand scale than a Socialist. In 1934 the Communists began the historic Long March, escaping encirclement by the Kuomintang. It was civil war, bloody and brutal.

In 1937, the year K-W headed towards Yunnan, the Japanese, now intent on conquering China, launched an attack on Chinese troops near the Marco Polo Bridge (Lukouch'iao) outside Peking. But the Japanese had misjudged the Chinese. By attacking a divided nation they had hoped for a swift victory, but following the attack Chiang Kai-shek called on the whole country to resist as one people. The call was answered. The Communists and Kuomintang came together in a common cause. The Russians, not without self-interest, financed the Chinese war machine. The Sino-

Japanese war effectively ended any really meaningful European influence in the country.

It was scarcely surprising that when K-W crossed the border between Burma and China he failed to receive a warm welcome. Customs officers went through his baggage with a fine toothcomb, demanding to know exactly what he was carrying, and for what purpose. Finally he was allowed to make a camp in the grounds of a temple where small boys were being trained for the Buddhist religious life. ' ... they chant sentences over and over again like learning formula. It reminds me of village board schools forty-five years ago,' he wrote in his journal.

The border area was controlled by Kuomintang forces and the private armies recruited by local merchants. K-W met a 30-year-old soldier who told him that he had already fought in four wars.

After six days of enforced idleness, under what was virtually house arrest, the senior official, the Hsien Kuan, and a schoolmaster who spoke fluent English arrived with an official memorandum declaring that as K-W was carrying arms he was now formally under open arrest, and was forbidden to proceed to Szemao. He was also forbidden to collect anything, or to take photographs, and the Hsien Kuan said he would prefer it if K-W moved closer to his quarters so that he could keep an eye on him while waiting for further instructions from Szemao.

'So the longer they had to think it over the more they thought up against us,' he wrote in his diary. 'It is all a ramp. They tremble for their jobs. China is going red all right with tape. We fumed, expostulated, but we had no documentary evidence to support our statements other than Rangoon passports, which they have taken from us. The day passed in gloomy silence.' (He had a European companion with him in China, whom he simply refers to as 'B'.) The days dragged by. He recorded in his journal:

> From time to time recalcitrant soldiers are beaten and at the second beat they would blubber like children and there was a good deal of chatty profanity between the beater and the beaten.

From the Thames to the High Himalaya

> A slipshod sentry stands outside the officers' temple. We are now definitely suspects and prisoners. If we stir out of the temple two soldiers are immediately told to follow us at a few yards' distance quite openly.

During the following days he attempted to see officials to plead his case, but was constantly put off. Finally, by refusing to go away, he met a Mr Li, 'a neat little man with a livid face, straggling hairs on chin and corner of mouth, rather wistful eyes peering through glasses. He wore a long blue gown, the conventional attire of China.' They drank tea and talked at length. It was all perfectly polite, but unhelpful.

Mr Li expressed an interest in science and asked K-W if he would take groups of schoolteachers botanising, which K-W readily agreed to do. At least it broke the monotony, and also enabled him to do some collecting. On these trips he gathered hypericum, *Jasminum arborescens*, *Quercus semiserrata*, *Asystana castinopsis*, *Carrusa*, and a variety of tea. One day he managed to shake off his guards and found an andromeda, lycopodium, some compositae, a grass, ferns, and a castonopsis.

On another occasion he managed to get as far as a nearby valley where he found stands of pure *Quercus serrata*, intermingled with a few *Q. griffithii*. There were willows covered with orchids, viscium, loranthus, ferns, and a gesnerid. The valley led into a gorge filled with lush jungle where there grew several different species of figs, a climbing milletha, euonymus, scutellaria, viola, stachys, tradescantia and polygonum.

With his movements so restricted, K-W's trips were rarities. Apart from the guards placed over him, the whole country round about was swarming with soldiers, many of them raw levies who did not know one end of a rifle from the other, and the heavily armed private armies of the merchants.

All his attempts to obtain permission to continue his journey failed, and finally he was forced to return to Rangoon, still hoping that he would eventually get official permission to travel to Western China. To this end he went to Singapore, stopping briefly at Port Swettenham in Penang where he collected mangrove flowers. In Singapore he took his case to the Chinese consul, who was unable to help. K-W's journal entry for 4 June 1937 reads: 'Heard from Chinese consul that I cannot go to Huepeh, so China is off. The cables to and from Nanking cost me another £5 nearly. Reason: unrest in Huepeh. Long explanations given that I cannot go because [I] did not comply with regulations regarding permit to botanise in Yunnan and Szechuan. Must evidently return to England.'

In the event he did not go home. His seed shareholders were happy that he should collect in a more accessible region. He had also become interested in studying the distribution of plants between Northern Burma, north of the Irrawaddy confluence, and the southern Malay Peninsula.

In addition he planned to seek out begonia species, a group of plants that he had taken little special note of in the past. This new interest may well have been inspired by the American Begonia Society, which since its inception in 1932 had been very active in introducing species from all over the world. After the Second World War K-W collected begonias for American enthusiasts.

From Singapore he travelled by boat to Calcutta, and from there went on to Rangoon, which he reached on 15 June. One of a swarm of fortune tellers who came on board told him, 'This very luckee years for you, Master. You very kind to everyone, everyone not kind to you. You get telegram this year, soon, veree luckee, making money, up to now not making any money.'

'Glib and ridiculous,' he commented in his diary. In fact the prediction was not so far from the truth. He received word that he had been given permission to go to Nam Tamai, but only if he undertook not to cross either the Tibetan or the Chinese border. 'What a load off my mind! It is like being given a ticket to heaven after wandering like a lost soul in purgatory for three months,' he wrote jubilantly.

While trekking into Northern Burma he developed a touching relationship, which was made the more poignant by the break-up of his marriage and the separation from his daughters. It was with the little daughter of his guide 'who cannot be more than ten, if that, who carried a kiddy-load ... Today she stuck close to me the whole day. I heard her singing the tune of 'Ould Lang Syne'! When very tired or hungry she whines and says so, almost weeping; the next minute she is laughing or singing a song. She has to *act* being tired, or cold or hungry, as a little animal would. She must express her real feelings.'

K-W was fifty-two, and beginning to feel his age. On 20 September, after a hard day trekking up the Adung Valley, he recorded: 'I find these marches far more tiring than I did in 1931. Still, with a little sustenance every four hours (and I have only two proper meals a day, and those frugal enough) I seem to get along all right.' The snacks were currants or a piece of chocolate. Indeed, his stores did not allow for over-eating. On 6 October he made this note on them:

> I am out of flour and have only a few biscuits left, and am living chiefly on rice and dahl; luckily I have tea, jam, currants, chocolate, pemmican and sugar, Quaker oats and Horlicks. Not in great quantity – but enough for ten days living economically. My routine now is into bed at nine with hot water bottle and a cup of Horlicks. Lights out at 9.30, up at 5.30, tea at six. I sleep fairly well.

In addition to the problems imposed by the situation in China, the expedition was dogged by accidents and troubles. On 14 October K-W slipped on a rock, was thrown down a slope and impaled through his armpit on a bamboo spike. 'Frightfully painful, but not serious. We went on past our previous camping ground, across the next torrent tremendously swollen and camped towards dusk. Place alive with sandflies. I was in such pain I went straight to bed with a cup of Horlicks. Later slept fitfully, light-headed, almost delirious at times. Could do nothing but pour iodine into the lacerated muscles.'

The following day he had three more falls, one over a cliff when he was saved from death by being caught by some bushes. On 29 October he was feeling particularly low. He had trouble with coolie strikes, and on this particular day they were so short of water that it was impossible to cook the rice ration: 'I felt ill and out of sorts, out of humour, out of pocket, out of everything,' he recorded.

There was a worse disaster on 11 November when he found the tent used to store seeds and specimens had been raided by local cattle: 'When I went to look at the rhododendron seeds in my tent this morning a scene of devastation greeted my eyes. Everything mixed up topsy-turvy, and trampled and chewed ... I was almost in tears. Months of work spoiled!' But after six hours of clearing up he was relieved to find that the damage was not nearly as serious as he had at first imagined, though the local headman added insult to injury by offering him two rupees in compensation.

With nothing to draw him back to England, K-W stayed on in the East to continue exploring and collecting. At the end of 1938 he took part in another American-backed joint zoological and botanical expedition in North Burma organised by Suydam Cutting and A. S. Vernay, and known, appropriately enough, as the Vernay-Cutting Expedition.

Apart from looking after the botanical side of things, he was also responsible for commissariat work. In November he wrote to Winifred from a tiny frontier village 108 miles from Myitkyina, where he was building and provisioning the base camp. 'Tomorrow I go down the road – up and down rather – a week's journey to meet the Americans, who are just about due in Myitkyina as I write.'

He was clearly enjoying setting up an expedition without having to worry about money. He told Winifred he was building the camp with 'the assistance of innumerable Lashis, Lisus, and undesirables'. His only regret was that when the American party arrived he would have to give up his comfortable hut with its huge fireplace, which was a special luxury as the nights were bitterly cold, to sleep in his tent 'which won't be funny. But I have *two* hug-wug-bugs.'

His self-imposed exile was beginning to work for him, and the deep bitterness and depressions of 1937 were disappearing. On 5 May 1938 he was able to write in his journal:

From the Thames to the High Himalaya

> I feel strangely happy and at peace tonight. Why? The reaction from depression and dark despair? Or because it was a fine day and I arrived not utterly exhausted? Or because it was a beautiful night, with a quarter moon shining overhead? Or because, as I believe, something greatly for my good has happened and communicated itself to me spiritually?

He was able to contemplate other things apart from the collapse of his marriage. Some of these were extremely quaint ideas, such as a scheme to turn papaya seeds into ersatz caviar:

> Papaya has a mass of sticky black seeds. One might set up a factory for making caviar for the general [public] by cultivating Papayas. One would have to soften them [the seeds] — a little dilute acid; and make them more mucilaginous — add a little weak Gum Arabic. And of course one would not be so dishonest as to call it caviar. It would have a trade name, which would suggest a select form of caviar ('specially selected') e.g. Curious Caviar, something much better than ordinary.

At this period he was also expressing concern about the harmful impact of man's activities on the environment: 'But man has permanently altered the vegetation of the globe and continues to do so at an ever-growing rate. There is far more destruction than conservation. How many square miles under tea, rubber, coffee etc.? Destruction of forest for pulp.' He was even concerned about the amount of land taken up for the burial of the dead: 'How many are buried and how many are burnt? What is the area of graveyards in Great Britain? If it takes seven years for a corpse to rot away to dust, how many offensive and dangerous remains are there?'

Despite being responsible for the day-to-day running of the expedition, he did find time to collect seeds of rhododendrons, nomocharis, berberis and iris, but he was not long enough in the area to have a proper seed harvest, and much of what he collected was in capsules that had survived the winter.

Frank Kingdon-Ward

The 1938 Vernay-Cutting Expedition was a happy experience. He had a great liking for Americans, and when he was invited to visit New York in 1939 to play a part in the World Fair he felt confident that at last he would receive the kind of recognition and generous backing that would enable him to mount much more ambitious expeditions, and, perhaps, see an end to his ever-present financial worries.

8

Plain Mr Ward

TOWARDS the end of 1938 K-W returned to England, and as usual when he was on leave he busied himself with lecturing, writing articles, and planning future expeditions. At the beginning of July 1939 he was away again, but this time to New York and the World's Fair, taking with him a collection of plants, including some rare nomocharis. In America he found that he was a celebrity. The rock garden at the Fair, which had been awarded a gold medal, had been dedicated to him. He presided over a gardening clinic for wealthy enthusiasts, and was lavishly entertained.

He records a dance at the Hunt Club in his diary: 'Jolly dance, very well run – band non-stop, too noisy, but rather good rhythm. I danced a lot with Carol ('Heart of Maryland') who was a real good sort.' He even got up onto the stage and showed the revellers how to dance the Lambeth Walk. There was also tennis and swimming, and nights out at expensive restaurants.

Wealthy gardeners clamoured for him to visit their estates and advise them. One of these estates was Nemours, the Wilmington, Delaware, home of the du Pont family. He discovered with boyish delight that he could call up seventy-eight different extensions in the mansion from his bedside telephone. They included his host's bathroom, the ping-pong room, the visiting maids' pressing room, the bachelors' house, the garden pergola, the boarding house, the radio room, the greenhouses and a variety of other apartments such as the Walnut Room and the Maple Room.

Frank Kingdon-Ward

K-W got tipsy on Old Fashioneds and a fashionable drink called Royal Shandy-Gaff, a mixture of champagne and burgundy, which was served in a silver loving cup. He also discussed plans for a major collecting trip for the New York Botanic Garden.

It was a glorious six weeks which, he recorded with evident satisfaction, only cost him $100, the equivalent then of about £20. But in August he set sail back to England and on the 24th stepped ashore at Southampton to a very different world – a dreary room in a Cromwell Road hotel run by people who were 'incompetent, slack and mean', and a country preparing for war.

When he called in at the Natural History Museum in South Kensington he found the place buzzing with activity. In the Botanical Department they were trying to sort out what should be taken to their warehouse at Tring, Hertfordshire, for safekeeping in the face of the threat of attack from Germany, or even an invasion. All over London there was the urgent activity of a nation taking to the barricades. K-W wrote in his diary: 'Tension rising to snapping point. London seems very empty, half the traffic has gone. Great activity, ARP [air-raid precaution], Auxiliary Fire Brigade, ambulances, barrage and so forth. Many signs of military activity, soldiers, lorries, despatch riders. I recovered my kit from the cloakroom at Victoria, which had not after all been destroyed by IRA bomb as I feared.' While he was in America he had read a newspaper story claiming that the entire railway station had been blown up.

K-W was now fifty-four. He had found New York exhilarating, and had started to lay plans for a grand American-financed expedition. But with Europe at war he knew that would be impossible. In 1914 his plans for a major Tibetan expedition had been ended by the declaration of the First World War, but then nobody believed it would last more than a few weeks; months at most. No such optimism was abroad in 1939. The conviction was that it would be a long and bitter struggle. It appeared to K-W that his career as a plant hunter and explorer, at least for the foreseeable future, was at an end. All he wanted now was to put his unique knowledge and experience of the Far East to some use in the war effort.

Plain Mr Ward

On 26 August 1939, K-W presented himself at the War Office with a letter of introduction to a brigadier. After being passed from hand to hand he finally met him. The brigadier asked him about the Yunnan Road, which K-W had not seen, but he suspected that the information the War Office had received was inaccurate. He wrote in his diary: 'I may get a job. Brigadier happened to be a keen gardener and wanted to talk shop when he discovered who I was; but we had no time for that. War Office crowded; war very much in the air.'

He did spend some time at the Natural History Museum helping to sort out the herbarium specimens to be evacuated. He also went down to Cleeve Court to see Pleione and Martha, and was told the house was to be filled with evacuees, some of whom had already arrived. He was pleased with the garden, which had been beautifully tended by Florinda and her gardeners, but his meeting with her was not a success. In his gloomy London hotel room, he noted in his diary: 'F very vituperative – jeered at me for scramming to USA. But as I scuttled back as soon as war seemed to be imminent I thought I might get a good mark for that. The scram was arranged three months earlier anyhow.'

He was at a loose end. Hating the dullness of his London quarters he roamed the streets, observing a great city preparing for war. In his hotel room he carefully wrote down his impressions with his usual dip pen: 'London is very quiet, waiting, waiting. Resolute too. I went to Hyde Park to see the AA guns. Balloons appearing in fresh places daily. Ten big excavators taking up position near Serpentine. Band playing but park very empty ... an unusual quiet reigns. Fewer people about, fewer cars on the streets, fewer buses, fewer trains. Men digging trenches all day ...' It was 27 August, just a week before Britain and France declared war on Nazi Germany, following Hitler's invasion of Poland.

On 31 August he lunched with one of his most influential backers, Lionel de Rothschild. The meeting was something of a wake. The day before they had been to the Royal Horticultural Society's monthly show at Vincent Square near Victoria, and had been depressed by the lacklustre quality of the normally superb Alpine Garden Society exhibit. Rothschild was worried that the

Frank Kingdon-Ward

anti-aircraft guns emplaced in his famous gardens and nursery at Exbury, would, if used, shatter the greenhouses. Over lunch the conversation was simply war.

On the day that the German forces marched into Poland, K-W watched the balloon barrage being raised over London. 'The silver sausages menacing against a stormy sky,' he wrote. 'That night the blackout began; henceforward London is in Darkness from sunset to sunrise. We are now waiting ... waiting. The work of protection goes on ceaselessly. Sandbagging, strips of paper across windows.' On Sunday, 3 September, he made a note that war was declared at 11 a.m. on the radio and twenty minutes later came the first air-raid warning.

> A warden I met asked me to get under cover as quickly as possible. The sinister warble of the sirens was a dreadful doom-like sound – one's heart almost stopped for an instant. After the all-clear, which came very soon, went and got a gas mask. In the afternoon to the park to see the AA guns and big excavators at work digging earth and gravel for sandbags. Many Special and Reserve Police and others at work. All the Special Police in tin hats now on duty – hundreds. A fine day after heavy rain and thunder last night. Balloons look like white yachts sailing a tranquil turquoise sea.

He went to look up some friends, but they were out, so he made his way to Oxford Street and walked to Piccadilly for tea, and from there to St James's Park and Horse Guards' Parade '... where activity prevailed. Soldiers entering and leaving the War Office. Despatch riders coming and going, cheers for the PM in Downing Street.' The excitement and activity all directed to an end that could be guessed at but not accurately predicted, was the stuff of expeditions. The exhilaration worked on K-W. The day after the declaration of war, he wrote: 'Britain getting into her stride. Streets full of war reserve police, ARP people, ambulance drivers, soldiers. Marvellous blackout. "Such night in England ne'er has been".'

Plain Mr Ward

At this time he was seeing a good deal of Florinda, but not with the hope of any kind of reconciliation.

> September 15: Saw Florinda by appointment. She made me mad as usual – and got mad herself. Always too plausible – a bad sign in man or woman. Now I am beginning to think she really does mean well, according to her lights – which aren't mine. We parted the best of friends in the rain – mainly, I suppose, because I had given her five guineas to buy a coat. After all I *had* the cash and she said she could not afford a winter coat; I had just bought myself one.

He also kept up a fairly regular correspondence with his daughter Martha, who was living at Cleeve Court. He loved to play with words to amuse her, such as in a letter about the blitz:

> As I write the battle goes on, but I'm not going to say anything more about bombs, because really they are becoming a bit boring. Every day someone tells you about their bombs, and so it goes on. People simply bombard themselves and it is easy to become a bomb bore and even a bit bombastic; in fact you bombboozle one another with the silly old bombs. For instance we were discussing a very tall building the other day – or what remained of it, and a woman said it must have been hit by incendiary bombs and I said no, by high explosive, and she said how do you know? and I said well look at it it must have been *very* high to reach the top of such a tall building. She puzzled over it a bit and then gave it up.

Later in the letter he describes how he was nearly hit by a bomb in High Street Kensington, and could not resist a school-masterish joke: 'Oh, dear, there I am being bomb boring again telling you about *my* bomb.'

It was at this time that he received a letter from Dr W. J. Robbins, Director of the New York Botanic Garden, saying that the Garden would sponsor an expedition. K-W wrote back asking for a postponement as he believed his first duty was to volunteer to serve in the war. However, his attempts to get a war job were repeatedly frustrated, due to his age. On 25 September he wrote:

> I am determined to make another effort to be taken on; dressed up and went to the War Office. It was crowded of course. An NCO at the door asked me who, or what, I wanted, and passed me over to a colonel in mufti, who took me outside. I asked for an interview with General Beaumont-Nesbitt, he said not without an appointment, asked me to write. I suggested the telephone was quicker, but he said a subaltern would answer and keep me away. The colonel was very nice and quite human, and showed a flicker of interest when I mentioned India.

K-W wrote the letter and three days later received a reply from Beaumont-Nesbitt saying there was nothing at the War Office for him.

Eventually he did get a job, once again in the Censors' Office, and he loathed the work then as much as he had during the First World War. In his view it was a job for the lowest form of human life. He expressed this opinion in a letter to Martha: 'Reading other people's private letters, or confidential business correspondence even, has always seemed to me to be the lowest form of animal life. Quite unicellular in fact, could go no lower ...'

A lady, then in her early twenties, who was also working as a censor, remembers meeting him.

> My memory of him is absolutely clear. In 1940 I was posted to telegraph censorship, which was in the Prudential Building in Holborn in London. One day I received a document in an Asian language and was told to take it to a man called Kingdon-Ward for translation.
>
> I went into this large room where there were dozens of people working at desks, and I asked for Kingdon-Ward. The man in charge said: 'You mean old Kingdom Come'. I was shown to his desk. I looked at all these people, all busy, and there was a rather small, shrunken, rather shrivelled little man, a bit of a mystery man, but very courteous. I handed him the document, and I asked him: 'Are you the

Plain Mr Ward

Frank Kingdon-Ward?' He smiled and said: 'I like the *The.*' He had no identity in the department where he worked, but he accepted his lowly role with great charm.

As much as anything to retain his sanity he occupied his spare time trying to write a play, and also commenting on the war. One rather extraordinary article was published in the *Daily Telegraph* in October 1940, in which he argued that the war was due to a resurgence of the Neanderthal strain still in the blood of the German nation, and also the Russians. Neanderthal man, he maintained, had not been wiped out by *Homo sapiens*; instead they had interbred, particularly in Northern Europe, where the Neanderthal character had survived in a tribal form:

> Daily the frantic utterances of the Nazi leaders prove it, for this war is a tribal war; the echo of more ancient tribal war. For ten years Germany and Russia have been pulling down the blinds, not openly on their own lands, but wherever their influence has reached. Biologists cannot fail to see in the twilight a sudden recrudescence of Neanderthal man.
>
> We cannot get away from the horrible truth that behind the immense glittering facade of Western culture lurks the black shadow of Neanderthal man. The cold ferocity, the snarling rage, the sterile sadism, the grunting sensuality we see all around in Central Europe today are, in the literal meaning of the words, brutal and inhuman.
>
> Blind and deaf to all chivalry which has flowered down the ages, Teuton and Slav, whose ideals are always animal, have proclaimed themselves with one voice the Paleolithic People.
>
> So it would seem that at long last man has taken up arms to renew the grim struggle he unwisely abandoned one hundred centuries ago and cleanse the earth of this foul taint.

Mercifully for K-W the tedious drudgery of the Censors' Office suddenly came to an end late in 1940 when he was ordered to Scotland for special training before being sent to the East. Although he retained his rank of captain, and had a military number, 00100, he travelled to the Far East as plain Mr Ward, botanist and plant collector. His real task was much more important, and at the time regarded as vital in view of the threat to the British Empire from the Japanese. He was to find routes for troops and supplies that could be used to link the Allied forces with those of the Chinese.

On 22 April 1941 he took a train from Euston to Liverpool and booked into the Adelphi Hotel, which he declared was 'much more depressing than a Yunnan caravanserie, or mule inn'. After waiting for a few days he boarded the MV *Imperial Star*, delayed as a result of being hit by a bomb which had pierced the deck and passed through the crew's quarters, but failed to explode.

By 10 May they reached Freetown in Sierra Leone to find a mass of international shipping both inside and outside the harbour. The town itself was filled with soldiers and sailors, and was crawling with spies. His view of the West African port was not flattering: '... niggers and mangoes, *not* inspiring'. Here the *Imperial Star* left the protection of her convoy and ploughed on through bad weather in the South Atlantic to Cape Town, where she anchored under the lee of the breakwater. K-W wrote enthusiastically: 'The lights of Cape Town are wonderful. It is the city of a dream – or is it just the novelty after eighteen months of darkness?'

On 27 May he left by train for Durban, crossing plateau country that reminded him of Tibet. The east coast city reminded him of 'Blackpool with the chill off it'. In a letter to Winifred he amended this to: '... Worthing with the chill off on a Bank Holiday after an unusual influx of nigger minstrels. The Zulu rickshaw runners are pure musical comedy.'

He stayed a week in Durban before boarding a flying boat for Lourenco Marques, now Maputo, in Mozambique. From there they flew to Mombasa, crossed part of Lake Victoria, which he observed was carpeted with white waterlilies, and up the Victoria Nile to Murchison Falls. They were only flying at about three hundred feet, and K-W was entranced by hundreds of hippo

Plain Mr Ward

'floating, swimming, blowing, playing water polo, it seemed', and by the crocodiles. 'When I could drag my gaze from the fascinating river I saw buffalo on the forested bank and hundreds of reed buck.'

When they reached the White Nile the sights and signs of war returned with camouflaged oil tanks and fleets of lorries on the banks. At Cairo K-W booked into The Continental, and had a good grumble:

> ... all hotels in Cairo are bad and expensive. The servants treat you with veiled insolence, they are intolerably lazy. I regard the Egyptians as one of the most debased people of the world. They seem to do nothing with their country, we protect them, while a minority of them make a fortune out of us, and the rest starve. After the war they will swagger about, tell us to get out, and make believe they won the war. The greasy, degraded civilisation.

The journey from Durban to Cairo, about four thousand miles, had taken four days. The next leg took K-W, and the senior service personnel making the same journey, over Sinai to Akaba, and thence to Basra '... across the white-hot sand. Then over terrible jagged, slag-like mountains, and almost due east over the top of Arabia. Flat salty desert, featureless until we reached the flooded Euphrates and date palms and huts of Marsh Arabs.'

Bahrain and Dubai followed, and a landing at what K-W called Jubami, but was probably Jiwani on the coast of Baluchistan. At Karachi he got into difficulties with the health authorities. 'I was lifted up with a pair of tongs and deftly deposited in the isolation hospital many miles out of Karachi on the edge of the great Indian desert. Thus it is to be an untouchable. My yellow fever immunity was not in order. I am a prisoner for 24 hours. Doctors, customs officers were all quite nice, but it is a bore.'

He went by train to Calcutta, on to Rangoon and Bangkok, and a spectacular crossing of the Indian Ocean to Penang. 'The sea was all the colours at the blue end of the spectrum and then some – violet, indigo, blue, blue-green, aquamarine, ultra-marine, turquoise, jacinth, jade, hyacinth, Prussian blue, true blue, little boy blue; and the clouds were smoked silver, white gold and platinum.'

At Singapore he had some difficulty with the immigration officers, who knew nothing of his mission, but he managed to pass himself off as a civilian botanist on government business. Social life in Singapore carried on as though there was no war, or threat of attack by the Japanese. He visited friends, danced, dined and drank, and waited with increasing impatience to be given the go-ahead for his mission. Security was so slack that he recorded: 'A good story is being told here of secret documents which were treated with fanatical veneration and marked out for destruction. Nevertheless the Admiral's white suit came back from the dhobi wrapped up in one lately.' Censorship was minimal and he was able to write to Winifred of military shipping and coastal defences.

In July he was in Thailand about to embark on his first official commission. He wrote to Winifred from a Chinese hotel in a Thailand township.

> It is a hot sticky night and I am dripping with sweat. The street outside is crowded with a motley throng of Siamese and Chinese, with a sprinkling of Malays. Bedlam. In the open shops shiny-faced coolies are shovelling hot rice into their mouths, or sipping tea; clerks are playing snooker on a faded green billiard table that probably never was sea-level; young bloods are having their hair cut. Along the muddy river are the masthead lights of junks, sanpans and other clumsy-looking craft; and a rich suety smell compounded of rotten fish, durian, sweat and open-air latrines pervades everything.

This first war expedition was to survey a route from Thailand to the Indian Ocean. He told Winifred: 'I am setting out early tomorrow morning to try and cross the peninsula to the Indian Ocean, mainly on elephants, though we do the first day's journey by boat up the river.' There were no roads, and it was in the middle of the rainy season. He expected it to take four days. It took six, but was completed without incident.

In October he was in Rangoon, staying at the Strand Hotel. He wrote to Winifred: 'I am off on an expedition plant hunting on the Burma Frontier, a pleasant change after three months in Singapore.' The paragraph was underlined with red, possibly a hint to his sister that plant hunting had now become a euphemism for his secret trips. Part of his work was for the military survey service, which was responsible for supplying the forces with reliable maps of the area threatened by Japan.

The following month he was in Raffles Hotel in Singapore, scarcely four weeks before the declaration of war with Japan. The threats from Tokyo still seemed to have little impact in Singapore, where military activity was more a formality than anything else. When he was setting off on a brief cruise among the islands, which appeared to have something to do with plans for the defence of the main island '... the sirens howled and a squadron of fighter 'planes roared across the waterfront. It was something like London – without the bombs! Realistic though, and I got a twinge of *the old thrill*. It was only our regular weekly practice,' he wrote to Winifred. When the aircraft disappeared from sight, 'The sky was all mutton-fat clouds with black bats' wings threatening donner blitzen and on the horizon puffs of cumulus towering twenty thousand feet into the sky, as though the rim of the world was splitting in all directions letting the gas out.'

War could not have seemed more remote as he sailed among the islands. 'The nearest islands were striped at the base with big bars of golden sand. Those in the middle distance caught by cloud shadows were indigo humps looking as though hacked out of cardboard and stuck in the sea. Those in the distance curled up at the ends, and seemed to be floating in the air.'

Quite how much useful work he and his companions did is debatable. They became so engrossed looking at the coral that the boat ran aground before 'we chugged home through still blue seas in which jellyfish as large as footballs idled and flying fish leaped'.

After a lazy weekend on a rubber estate on the mainland of Johore, K-W was on the move again: 'I am on my way to Burma. You will have seen from the papers that the Japanese are again threatening the Burma Road.' The day before sailing he had spent his time studying maps of Burma and Indo-China. What was vital was that if the Japanese did take the Burma Road there would be an alternative route available to maintain contact with the Chinese forces in Kunming.

There were the usual hold-ups and he was delayed for a while in Rangoon, where he spent one dismal evening at Maxim's Dance Hall. 'What a joint. A bad dozen nondescript-to-ugly dames, mostly fat, coffee-coloured, and generally unattractive. One Burmese woman, well past her youth, was — passable at a distance.'

From Rangoon he went up to Pegu, where the ditches were filled with scarlet waterlilies and pink lotus, and the rice was ripening in the paddy fields. On 1 December, with war with Japan only a week away, he was still in Pegu, '... an unsavoury hotch-potch of huts, shanties, shacks, shops, dirty, crowded and smelly'. The tension and excitement over the Far East situation were growing daily: 'Strips of army lorries passing through on the way to the front. An oppressive atmosphere as though anything might happen any minute.'

On 7 December the Japanese forces landed on the coast of Malaya, and bombed Singapore and Hong Kong. The following day Winston Churchill penned an elegant letter to the Japanese Ambassador in London announcing that a state of war existed between their two countries. He signed the letter: 'I have the honour to be, with high consideration, Sir, Your obedient servant, Winston S. Churchill'. In his history of the Second World War, Churchill observed: 'Some people did not like this ceremonial style. But after all when you have to kill a man it costs nothing to be polite.'

There were now all the signs of a real war. Troops marching single file, mules loaded with tommy guns and ammunition, lorries and bullock carts. 'The war is becoming something which distinctly affects every man, woman and child, not something you listen to on a wireless or read about in the newspapers. It does not just inconvenience our lives, make us deviate from a routine, deprive us of luxuries, may deprive us of necessities [sic], cost us our lives. It is really a world war.'

As the conflict spread through the East K-W declared that he had ceased to take any interest in botany. The reality of war bore heavily upon him, and on Christmas Eve, when he had stopped

Plain Mr Ward

for the night in a small village, he lit four candles outside his hut, and prayed. He also wrote in his journal:

> This is not a war of conquest and exploitation, but of conquest and extirpation. A war of survival and replacement. The Japs in Malaya would not be like the British, a handful of administrators, traders and planters living among the native population. They would themselves quickly become the native population by the simple process of exterminating and replacing the earlier population. The Germans have admitted that this is their aim and their actions do not belie them. The Japanese tend to do it tacitly, but what Malaya will be like a few years hence one boggles to think. A vast reshuffle is going on, a hell's broth mixture of Malay and Chinese, Indian, European and Japanese. It may take a thousand or ten thousand years for a definite breed to emerge from this kaleidoscope of colour.

His gloom was justified. The war with Japan was going badly. The sinking of the *Repulse* and the *Prince of Wales* was a staggering blow, as were the successful Japanese landings in Malaya, and the devastating attacks on Allied airfields. By Christmas Eve it was clear that Hong Kong could no longer hold out, and Penang had been invaded. Singapore was under threat. On 15 February it surrendered unconditionally to the Japanese. In the same month Japan invaded Burma. Over Christmas 1941 K-W was on the Mekong River, searching for suitable crossing points and persuading the local people to hide their boats, not to destroy them as they were doing, because they could be of use to Allied troops.

On 29 December he was at Mong Yang close to the border between Burma and Indo-China, where a few days later he met up with part of the Chinese army. He had recovered his appetite for botany and had gathered a hundred species of plants at Mong Yang, although he makes no mention of what actually happened to them.

As well as surveying and photographing the Burma–Indo-China section of the Mekong, it seems likely he was engaged in some kind of liaison work with the Chinese forces in the area, and his journal contains observations about their strength, and the siting of their defences. On the whole he was impressed by their efficiency.

By 7 February he was back in Rangoon, now a sadly changed city. 'Very different from the usual shouting, jostling crowd. Few people, subdued. Taxis hardly available. Asking five rupees to do two miles or less! I went to a dreadful place called Devon Court, had a bath and a shave.' K-W had no particular love for Rangoon, and his stay was made worse by the behaviour of the military. He was given only a sketchy outline of what he was supposed to do next, and nobody seemed interested in the slightest in what he had already done. He thought the whole thing a farce, and that he was only being given jobs in order to humour him. He was cast into a bout of depression. '... it all seems wasted effort. The work done useless, the job assigned just to keep me employed. Life here is hateful. The house servants robbed me the second night I was here of 75 rupees, stolen from my pocket book, and that upsets me a lot.'

On 19 February he left Rangoon for India on what turned out to be a refugee ship:

> Crowds and crowds fleeing to India ... The crowd of evacuees milling around was awful. One could hardly move for weeping women, piles of luggage, babies. Actually we got sorted out pretty quick, and the ship lost no time in getting underway, not, however, before a woman had dropped her small child into the river – it was fished out none the worse – and a number of coolies who had no tickets had been thrown off the ship. A strong force of police lashed around with their canes, driving away frantic people who wished to flee from the battle front, but had perforce to flee from blows rained on them.

When K-W arrived in New Delhi he found there was nobody to greet him; in fact the authorities had not even been informed that he was on his way. He came with a plan to transport men and supplies into China from India by the Rima Road. This was almost immediately rejected in favour of a more difficult route, then being negotiated with the Tibetan Government – 'To use the Gylam to pass stuff into China! And it is a 1,200-mile trip from Gangtok, over the greatest ranges of mountains in the world!' he exclaimed. But on this occasion he was not too depressed by what he saw as an ill-informed rejection of his ideas, and busied himself writing reports on possible overland routes to China. He also spent time planning his next journey, which had been switched to North Burma from a projected survey of the Rima Road.

From New Delhi he went to Calcutta, and in March 1942 left that city for the Brahmaputra, which was being used along its navigable length to convey war supplies. From there he travelled to Imphal in Manipur and on to Patel, where he was to meet up with a man called Hesketh. Once again his commission was to search out supply routes as well as escape routes for refugees. In fact he arrived in Manipur in the midst of a massive exodus. He recorded:

> We found a great trek in full swing. Camps crowded with Indians of all sorts. Long files trudging patiently along in the dust and heat with bundles and boxes on their heads, or babies on their backs. Parties sitting under the bushes trying to shelter from the almost vertical sun, cooking meals, washing their clothes in streams. Occasionally a bearded Sikh on a brand new bicycle, his kit tied on behind, came along: then the endless procession of walkers was resumed, parties of 20 or so, 50, over 100, the vast majority men, but some were women. 'We are not running away from the Japs, Sahib, or from the bombs, we are running away from the cruel Burmese who will kill us.'
>
> Very occasionally a car jammed to capacity with men, women and children and bundles, or a bus, which had no right to be there since the Government has commandeered them, passed. Meanwhile we passed derelict cars, an occasional corpse, a black naked body scarcely visible against the charred undergrowth. Still the endless queue of coolies in soiled white clothes tramped on raising clouds of powdery white dust, thirsty, footsore, weary, strings of

Burmese bullock carts jammed to the roof passed. We saw big camps with cooking fires, carts massed, bullocks, men and women stretched under sheets like shrouds.

The war in Burma has brought out the underlying character of three communities. The British, up against it, are at their best, working like slaves, devising, sacrificing, helping, improvising, carrying on. The Indians are helpless, the rich selfish and indifferent, interested only in their families and goods, the poor patient, uncomplaining (except when ordered otherwise) and entirely helpless. The Burmese cruel ...

Black-marketeering and profiteering were rife as the refugees were fleeced in their desperate desire to reach the safety of India. K-W looked out on the beautiful Manipur landscape as he wrote in his journal: 'Never again will the slumbering hills be quite the same.'

Eventually he reached Mandalay, which three days before had suffered a massive air attack. The railway station was in ruins, the charred corpses of horses littered the streets, nearly every building was gutted. It was attacked again before he left and was still burning when he set out for Putao. It was there on 9 May that he received word that Myitkyina, Sumpra Bum and Utal were to be evacuated.

The position in Burma was dire, but instead of making a swift and safe return to India, he decided to continue with his search for safe military and refugee routes. His plans were dramatically changed when a letter was dropped at his base at Putao telling him that a large party of Indian and English refugees were on their way from Myitkyina, and that he was to prepare a landing strip so that RAF transports could airlift them to safety.

K-W complained in his journal: 'Government must be crazy to let refugees start a 220-mile march to Putao, there is neither food nor shelter, even if we can prepare the landing ground.' In a clear reference to the problems he had with both military and civilian authorities, he added, quite spikily: 'They won't listen to their experts when there is time to prepare, and then ask them to perform miracles. If Indian refugees in numbers come up this road, the rains just starting, and if women and children also come, there will be starvation, disease, fighting between the Kachins, Shans and Indians, and everybody here will be in deadly peril.'

But despite his misgivings he got together fifty coolies, found a suitable site, cleared it of stones and tree stumps, and cut the grass. He constructed a remarkable wind indicator from a butterfly net and a strip of dhoti rigged on a thirty-foot bamboo pole. With the airfield completed he left to find his way back to India. His search for safe routes for troops and refugees had to be abandoned.

K-W was nearly fifty-seven, and the years of strain and travel were telling. When he finally got away from Putao he wrote that he could have fallen on his knees and thanked God. 'All afternoon I felt I wanted to cry, but whether with relief at getting away, or because I was frightened of what lies ahead of me, I don't really know. After the last few days of alarm and despondency I felt confidence returning.'

He had feared that the Japanese would occupy Putao before he could leave, but now that he was heading for India he believed he would be able to escape a Japanese prisoner-of-war camp. However he was unsure which route to take. He was running desperately short of money, which he needed to hire some porters, and was finding it almost impossible to communicate with the tribes. Ideally he wanted to get into Tibet and reach India via Gyanda or the Tsangpo Valley, which was country he knew well. 'Under the circumstances,' he wrote in his journal, 'it is very difficult. China and Tibet are both taboo – but in a crisis that hardly counts.'

The route he finally settled on was through the Diphuk Pass to Rima, just in China, and from there to Sadiya in Assam. The trek took nearly two months, and K-W was back in Calcutta on 25 July. On 9 September he was ordered to New Delhi to see General Wavell. He described their meeting:

> A spruce, thickset, sturdy man with small twinkling eyes, a military moustache, a lined face, scholarly brow, steel-grey hair. He shook hands warmly, was, like all really great men, courteous, without affability. We walked over to a

map. He asked me where I had been, and about roads. When I told him he gave a few seconds' thought and said 'I see' shortly, and one felt certain he did see everything, and every implication behind your remark.

K-W was delighted with the interview. For the first time since the day he left the grinding tedium of the Censors' Office in the Prudential Building, he felt that he was being taken seriously.

No doubt Wavell did take K-W's information into account, but the interview did not lead to a new post, although K-W still believed he could and should play an important role in planning the campaign against the Japanese, relying on his unique knowledge of Assam, Burma and the Sino-Tibetan border country. Few men knew more about the Triangle, the beautiful, treacherous wilderness bounded by the Mali Hka and the Nmai Hka, the rivers which flowed into the great Irrawaddy, and the area which was to become the theatre for some of the bloodiest fighting in the Burma Campaign.

The job that K-W did get was to establish dumps of stores and munitions in readiness for the planned assault on central Burma. In February 1943 he was in a camp in Manipur on the Burma frontier. In a letter to Winifred, he wrote: 'I am engaged on a smuggling racket. More accurately perhaps described as running the blockade! A hard life! ... I find it hard work slogging over these mountains by rough tracks at the age of 57. Having done it all my life, it is almost second nature to me, and I can still do it — but it's beginning to tell, and will be worse in the rains. I am fairly well paid by the Government however, so it's worth it.'

The weather was really a more serious threat than any Japanese patrols. After a storm one of his dumps was swept away by a river which rose three feet in twenty-four hours. Worse even than the weather and the enemy was the greed of the contracting companies he had to deal with. He complained to Winifred:

> I have been having bitter fights with Indian contractors. Talk about profiteering! They don't begin to consider a job of work till they see 200% profit! I've twisted the tail of one and he's fairly howling — caught him nicely. But it's

> an uphill game, and one sees the dirty side of business in this racket. I employ 50 persons and with a few exceptions they are a pretty unsavoury crew. They include contractors, boatmen, interpreters, clerks, storekeepers, coolies, and others, besides personal servants. Several of them I like very much, others are naughty children, others just greedy pigs. I have to do the best I can with them, and I am also father confessor, confidential adviser, doctor, and of course – paymaster! That's what counts.

The majority of K-W's letters to Winifred were full of rich description, sometimes self-revelation; in a way they were extensions or at least additions to his private diaries and journals. A letter written to her from his base camp in Manipur at the beginning of April 1943 brings alive the way he was living at the time: 'I am still hard at work on the frontier, getting about a good deal, sometimes a trip of 100 miles down the river to Silchar, by country boat – a slow business, but much slower coming back, sometimes a trip up the Tuivai, through the rapids by dugout, sometimes several days' march across the mountains.' The weather was intensely hot, and with it came

> all the maddening birds like barbets and cuckoos (*our* cuckoos, which don't call cuckoo) are playing the overture, tonk tonk tonk (coppersmith), took-a-rook-took-a-rook-a took-a-rook ad. lib. (another kind of barbet) and dozens more. Enough to give one softening of the brain. However one gets used to it – like the sandflies. Apart from this, life though hard is pleasant enough, certainly not monotonous in the absence of any attempt at routine; in fact each day is improvised (mean the work is) as it comes along; and only when I am at my base camp on the river, a few days at a time perhaps twice a month, is there anything approaching routine. There I live in one bamboo hut, a few hundred feet above the river, and my assistant lives in another across the river and about $\frac{1}{2}$ a mile distant, where the main camp and the *godowns* [warehouses] are situated. Every day I visit the camp, discuss the position with Mundy

(my invaluable assistant), inspect the *godowns*, write out the order (if any), and return to my *basha* on the hill.

It was not exactly an exciting life, but he was content, and he was left with a good deal of time for writing and thinking. He was still managing to add to his income by producing articles for magazines such as *Blackwood's*, which was a help as the financial demands from home were growing, particularly the school fees for his daughters.

For K-W what was important about the job he was doing was that he was making a positive contribution to the war effort, and was also out of everyday contact with Calcutta and New Delhi, with their irritations, unresponsive military men, and trying government officials. He was back in a jungle, alive, as he wrote, with 'the soft warblings of doves, the staccato bark of muntjak, defiant unto death, the grating screech of parakeets on the wing, the abominable jabberings and squealing of monkeys, the clipped cock-a-doodle-doo of the swaggering jungle cock, the joyous whooping of gibbons – what noises are sweeter!'

While much of his wartime travelling, searching for suitable dumping sites for military stores away from the villages which were often bombed by the Japanese, was done in the way he knew best – on foot – he did sometimes have a jeep at his disposal. Of one jeep trip during a brief return to Imphal, he recorded:

> We crossed two high ranges and very exciting it was. We climbed 2,000 feet in 20 minutes, round hairpin bends. Then we rushed up over the edge of the world, and looking down saw space below you [sic], all starry with flowering trees. Then you plunged down towards it round hairpin bends, the jeep turning on its axis like a wasp impaled on a pin. So we ran down the short wavelengths of road, with one wheel on the brink of the pit and the other trying to climb up the bank, and there was no time to be afraid. It was like riding forked lightning. My Indian driver was good.

Plain Mr Ward

Another of his jeep drivers at Imphal was 'a very fair-haired Northumbrian boy, rather shy, first-rate driver'. When he discovered it was the young soldier's twenty-second birthday he took him to the cinema in the city as a treat, and out to supper.

In May 1943, from a camp on the Assam border where he was waiting for a fresh assignment, he wrote to Winifred of the amount of trekking he had done in 1943.

> I have tramped 500 miles up and down these blessed hills — been some distance into Burma — since January, walked through two pairs of boots and one pair of shoes, and my last pair of pyjamas is — are — hanging in tatters on me. However as I paid 7s. 6d. for them at Marks & Spencer in 1939 they have done their bit, especially as innumerable *dhobies* have in their turn tried to break stones on them. We are not rationed for clothing, and can buy what we like, which sounds fine until you come to the three qualifying statements — if you can afford it, if it exists, if you know where to buy it. I certainly cannot buy clothes here, and if I stay much longer am likely to be reduced to wearing leaves, like a cocoon. There are plenty of fig trees.

In July he was recalled to Calcutta and told that he was to go to Poona, where he had been appointed to the RAF training school to teach airmen how to survive in the jungle. He arrived in Poona on 17 July and was lodged in a mess with a large number of young RAF men from Britain, Canada and New Zealand. Much to his disgust he was expected to sleep in a dormitory with four other men. He wrote in his journal: 'I do long for a little privacy. Two or even three is bearable, but five!' His job was to deliver lectures on the geography and jungles of Burma as part of a survival course for pilots and aircrew. Although he found the young airmen agreeable, their attitude towards India annoyed him. He said they had come out with the fixed intention of hating the country, and thought of nothing but getting back home. He complained that they had 'an infinite capacity for doing nothing'.

At the end of November he had an accident which could easily have been fatal, or resulted in injuries leaving him crippled for life. He was out driving in a jeep with two service friends. They were chasing a peafowl along a steep rocky mountain road when a hyena ran into the path of the vehicle. 'We all got a bit excited,' he wrote to Winifred.

> The driver trod on it (accelerator not hyena), and we beetled after the brute, which was now galloping up the narrow path. I foolishly stood up. The jeep hit the bank at about 30 m.p.h., stopped dead; I didn't. I just took a dive over the windscreen, landed on my face a dozen feet in front of the jeep, made a mess of my face, broke my neck – at least I fractured a cervical process. I thought I'd had it as they say in the RAF, but after a stunned silence of some seconds found I could just move my limbs, so I had not broken my back after all.

It was half an hour before he could be got back into the jeep and driven to hospital, where he stayed until New Year's Eve. In fact he rather enjoyed his stay in hospital, particularly Christmas: 'As for me I had a delightful Christmas in hospital. I was a walking case by then and went round to my friends in the wards, sat out in the garden under a blue sky and hot sun, drank a glass of sherry (so-called) with the sisters, wrote.'

He wasn't so pleased to be discharged. 'New Year's Day was brilliant as usual, but a flop for me. The mess was empty, recovering from a dreadful debauch, my arms were crippled, otherwise I was well and *looked* well (you should have seen my face the day after the smash – cat's meat!).' Although he made a remarkable recovery, the accident left him with recurring back pains for the remainder of his life.

In January 1944 K-W moved to Bombay on what he described as special duty, and enjoyed having his own quarters again; a real house, with a large room and bathroom of his own –

> rather a change from the tents, bashas, hospitals, dormitories, and similar places I have been occupying for nearly two years. True the house is an Indian house, hence the window frames don't fit and the monsoon fairly squirts

> through, flooding the floor; the two electric lights are in the wrong place; the walls, though thick, let the rain through so that a huge wet patch is slowly spreading over one wall inside and will presently support a crop of mushrooms – and other structural defects, to say nothing of design. But why worry – we shall be lucky if it doesn't collapse one day like a pack of cards.

The special assignment he had been given turned out, in his view, to be quite pointless. He wrote in his journal: 'This is a farcical job as far as doing any useful work is concerned. I am being paid for doing nothing.' So it came as a relief in October 1944 when he was sent to the Jungle Training School at Mahabaleshwar. At least at the school he was able to get out into the open, leading small groups of RAF men on tough survival treks.

The only compensation in Bombay had been having the time to catch up with mail from home, including long, erudite letters about music from his youngest daughter, Martha. He replied in a less than musically erudite way:

> Well, I don't hear much music, so it isn't much use trying to talk to you about your favourite subject. Besides, you know so much more about it that I should merely put my foot in the oboe if I so much as began to whisper about it. As for the flute, I don't give a hoot, I'd rather the lute, which I think rather cute. I do like the 'cello, for it doesn't bellow, and injure a fellow, by turning him yellow. In fact it is mellow. Well that's all I have to say about music for the moment ...

At the end of 1944 his work with the RAF came to an end, and early the following year he was sent to Assam to organise the evacuation of some tribesmen. In the March he was in the running for an intelligence job that would have taken him to Tibet, but to his bitter disappointment it fell through. In the entry in his journal for 10 to 12 April 1945 he wrote: 'Sometimes I feel desperate and want to kill myself. Sometimes a curious elation seizes me and I feel that all is working out right. But the chances I seem to have missed. It makes me shudder and lose faith in myself. Saps all resolution.' And in June he wrote: 'No news of a job, and few letters. It looks as though I am finished.'

Later that month he left the services and travelled to Assam to take up a job on a tea plantation.

9

Of Tea, and 'Planes, and Lilies

FOR most people 1945 was a year of triumph and rejoicing, but not for K-W. He was nearly sixty, an age when men are normally planning their retirement. What was he to retire to? A bedsitter in the Earl's Court Road in London, and a little part-time work in the herbariums of Kew Gardens and the Natural History Museum? Perhaps a few broadcasts and articles on old adventures until people grew tired of them? There was only one thing he wanted to do and that was to return to the wilderness to search for new plants; to follow trails which no other European had trod.

But the world was only just emerging from the shadow of war, and plant hunting and journeys of exploration were far down both national and private lists of priorities. K-W realised this and when he was offered a job helping to run the Khowang Tea Estate in Assam — it covered about a square mile and supported 1,500,000 tea bushes — he decided to take it.

The job was not what he wanted, but at the same time it was a relief to have an income and a comfortable base. He noted in his journal on 25 June 1945: 'Saw my bungalow. It is quite nice — I think I shall like it here for a bit — work dull perhaps, but I *can* do it, and shan't be over-worked. This bungalow — old-fashioned type — is very comfortable, even luxurious. Youngsters coming to a garden like this say "I think I would like to be a planter after the war".'

By 28 June K-W's mood had changed: 'Fine sunny day. I

can't work or do anything. Insects at night are a pest. I am through.' The following day his cook and houseboy walked out on him.

He got some relief from the irksomeness of a regular job when he was asked by the Tocklai Tea Research Station at Cinnamara in Assam to make a short trip into the Khasi Hills to look for wild tea plants that could be used in breeding programmes.

Often at the end of the day's work he would get out his bicycle and ride to a place where he could get a clear view of the Himalayas and 'clouds in tumbled tufts lying about pink in the setting sun. Spacious views, jungles, rice fields, tea gardens.' He was like a man exiled who can just see in the distance his now forbidden homeland.

On expeditions he had always been able to communicate with his servants and porters, but on the tea estate, where he did not speak what he called 'the coolie bat', he felt isolated from his work force. Like many men before him who found themselves in a similar position, he considered taking a native mistress. In the journal he kept while on the estate, he wrote:

> I can't fathom out their customs. A smile from a girl (who may be only fifteen) seems to be an invitation to sleep with her. If you don't take her at her word – or sign – and how can you if you can't speak the bat well enough to make a date! – after a few days she won't take any further notice. As a matter of fact the evenings are beginning to get a bit lonely and I wouldn't mind a girl in once or twice a week.

Instead of having a mistress, he shared his somewhat bleak existence with a tabby cat, four hens, two ducks and four pigeons, along with assorted lizards and frogs. The fowls were not a permanent feature as they were destined 'to be potted, a pet in a pot; not pitied'.

His dealings with his labour force were far from romantic. Apart from the day-to-day running of the tea garden he also had to play the part of judge and jury in minor domestic upheavals, with which he usually coped with good humour. He also amused himself restoring the overgrown garden in which his bungalow stood. He wrote to Winifred: 'I was very drastic about the garden when I came, and slashed around good and hard, uprooted and repotted the begonias, gave the ferns an Eton crop (with excellent results) and planted morning glories on the west side of the veranda.'

A letter written in August 1945, less than two weeks after the surrender of Japan, found him in an unsanguine mood, however.

> Yes indeed, what now! I'm afraid I'm a confirmed pessimist. I don't see much change, except in the material world: and I am enough of a scientist to know that all the talk about a change in human nature is just talk, and dangerous talk. Human nature does not change in 5 years, or 50, or 500 or 5,000 unfortunately. So long as it is human. I guess we've got to try and be a little *less* human.
>
> On the sociological side there seem to be two ways by which we can bridge the colossal gap (much more apparent out here than in Britain) between rich and poor, between those who have almost everything they want (except contentment) and those who really have not enough to eat. One way is to raise the standard of living of the masses; the other is to reduce the standard of living of the too-rich. The latter is the war way — but I'm afraid it won't be kept up and followed up.

Despite the uncertainty of his future, he admitted in a journal entry: 'If not actively happy here I am at least contented, and have stopped worrying over what can't be helped. Probably I am not "dynamic" enough to get the things I want.'

His tea-planting life was not all loneliness and work. There were the occasional trips to the country club that served the estate owners and managers. On Christmas Day 1945 he played the part of Father Christmas at the club, as he recorded in his journal:

> My fifth consecutive Christmas in the East. Went to the Club in the afternoon where I dressed up. Rode in a tank from Club entrance right across the polo field. I popped out — Everyone on the veranda clapping. Marched in all

doubled up with sack, children delighted (two cried with fright). Gave away the presents, and then more rides in the tank. Pronounced a success. I quite enjoyed it. Back to Khowang to dinner party. We played silly games. Got back about 1 a.m. Have not worn evening togs for nearly a year.

A correspondence which brought him increasing pleasure was with a young woman, Jean Macklin, the daughter of a distinguished Bombay High Court judge, Sir Albert Sortain Romer Macklin. K-W and she had first met at a lunch party in Bombay in 1944. Jean had had a chance conversation with a businessman friend who told her he was going to Sunday lunch to meet an explorer called Frank Kingdon-Ward. She recalled: 'I said to him, "Oh, lucky you, I wish I could meet an explorer!" He arranged for an invitation for me.'

She had been fascinated by exploration even as a teenager, though she had never heard of K-W. She was placed next to him at lunch, and they hit it off immediately. She was only twenty-three, thirty-six years his junior, but she felt no sense of being humoured or talked down to. He treated her as an equal, and she found him a quiet, reserved, and very modest man. K-W was certainly impressed by Jean, and towards the end of 1944 put her up for Fellowship of the Royal Geographical Society, describing her as a highly intelligent young woman. After that lunch they met only once more, at her parent's home, before she returned to England with her family in May 1945. Jean and K-W began a regular correspondence, at first chatty letters about what they were doing, but gradually becoming warmer and more intimate.

On 1 January 1946 the future for K-W again looked bleak. His contract with the tea garden was at an end, and his entry in his journal for New Year's Day read: 'My first day of unemployment – a bad start for the new year.' In fact his unemployment was short-lived. Early in the year he was asked by the American airforce if he would go to Manipur and search the forests, gorges and mountains for US aircraft which had crashed during the war, and recover the remains of the crews so that they could be decently buried. It was a grim task but one that returned him to the wilds, and proved to be a decisive step back to plant hunting.

Based in Imphal, he scoured the hillsides and gorges of Manipur for the wrecked aircraft; he also botanised as he went along. He was particularly intrigued by one peak, Sirhoi, rising 8,500 feet in the divide between Assam and Burma. Instinct and experience told him that he would find good plants there. Although there was no evidence to suggest that any 'planes had crashed on the mountain, he insisted that it should be searched.

He and the two GIs accompanying him made part of the journey to the mountain by jeep, and the final stage on foot. By the time they reached a suitable camping site the young soldiers were exhausted, and happily agreed to a rest day the following morning.

K-W climbed the mountain with a local native guide. There were no aircraft, but there were plants, and one in particular aroused his curiosity. It was not in flower, but it appeared to be a nomocharis, which was especially exciting as there were no recorded nomocharis nearer than the other side of Burma. The dry stalks suggested that it grew from six to ten inches tall. The bulbs were quite large with pale yellow scales, and K-W guessed that the flowers would be white, although his guide said they were magenta.

K-W collected the few seed capsules to be found, dug up fourteen bulbs, and three rooted pieces of a crested iris he also discovered. He sent half the nomocharis to the British political agent in Imphal, Mr C. Gimson, along with the irises, and the remaining bulbs to Mrs L. G. Holder in Shillong in Assam. Both were skilful gardeners. The seeds and capsules he sent to the Royal Horticultural Society in London.

First reports on the nomocharis, which K-W soon began to refer to as a lily, were discouraging. Both Gimson and Mrs Holder, after flowering their bulbs, described the colour as dirty white. Those raised from seed by the RHS, which flowered in 1947, were also written off with the same description. But K-W was confident that he had found a first-class garden plant, even if these were disappointing specimens of it. His confidence was not to be vindicated for another two years.

His work for the American airforce came to an end in March, and once again K-W was out of work; however, he kept busy with writing, mainly articles for the *Gardeners' Chronicle* and the *Royal Geographical Society Journal*. In May he received a cable offering him the editorship of the *Gardeners' Chronicle* at a salary of £1,000 a year. 'Very difficult,' he wrote in his journal. 'I have certain responsibilities, so it would be wrong to turn it down flatly, just because I don't like office work, and want to explore. Wrong also just to accept just because it is a safe job, and would end financial worries.' In two years' time Martha would be twenty-one and he would no longer be expected to help to support her. He decided on a compromise. He would take the job for two years. His suggestion was rejected.

Meanwhile his correspondence flourished with Jean, who was now working in Brussels for the Inter-Allied Reparation Agency. So much so that in July 1946 he noted: 'Jean and I are as good as engaged I believe.' And a few days later: 'I think of Jean more and more, and I miss her, I don't know why, I have only met her twice. Yet I seem to have known her for years. Am I in love with her I wonder?'

In August K-W was in the Shillong district of Assam searching for wild tea species for the Indian Tea Association when he was suddenly taken ill. The pain and the fact that he appeared to be suffering from internal bleeding led him to believe he was dying, so he listed his meagre assets and made a will. He was admitted to the Welsh Mission Hospital at Shillong, where a prostate was diagnosed. It took three operations to cure the problem, and the treatment swallowed up most of his savings from the ITA. While he was in hospital he wrote in his journal: 'If Jean were here how happy I would be.'

By October he was fit enough to resume the wild tea hunt, and was also cautiously optimistic about the future after receiving a letter from Dr W. J. Robbins, Director of the New York Botanic Garden, who had asked him to undertake a collecting expedition for the Garden at the beginning of the war. K-W had refused then because of the war, but now Robbins wanted to know if he would be free to mount the expedition if the trustees of the Garden were prepared to put up some of the money.

K-W realised that if he was going to conduct an important expedition it should be organised from England, and that he must return home, and renew his contacts with pre-war backers, Kew and the RHS. It took some time to organise a passage home, and he did not leave India until May 1947. He was taken ill on the voyage. By the time the vessel reached Port Said he was suffering from an internal haemorrhage, and was rushed ashore, but after hospital treatment was allowed to continue his journey. When he did arrive in England he was a sick man, and had to be nursed back to health by Winifred.

Once on his feet again he renewed his connection with Kew and the Royal Horticultural Society, and with famous nurseries like Ingwersen's, the alpine specialists at West Hoathly near East Grinstead in Sussex, and the lily and rhododendron specialists, Wallaces, which then ran a superb nursery laid out like a private garden at Tunbridge Wells. He found there was little enthusiasm for plant-hunting expeditions among amateur gardeners, who were tending to spend their money on restoring their gardens after the neglect of the war years. However, with the bulk of the money for the 1948 expedition already promised by the New York Botanic Garden, his financial worries were slight. He had also agreed to share the expedition with the Indian Botanical Survey.

More important than meetings with potential backers was one with Jean. Only the third since they first met, it took place at the Royal Geographical Society. The long correspondence had removed any sense of strangeness or shyness, and they agreed unhesitatingly that they must marry. The news was greeted with grave misgivings by Jean's parents, Sir Albert and Lady Macklin. They were alarmed by the great age difference, and the fact that he had been divorced, was poor, and was engaged in one of the most precarious of professions. They even asked K-W's first wife Florinda to try to influence Jean against going ahead with the marriage. Despite the opposition K-W and Jean were married in Chelsea on 12 November 1947, and before the end of the month were on board the SS *Franconia* heading for India and a collecting expedition in the Indian state of Manipur.

Not only was their marriage a wonderful partnership, it was also a genuine love match. A letter written by Jean to K-W only three weeks into the Manipur expedition is evidence enough:

> Darlingest Frank,
> Is it very silly to write to someone who is only in the next room? And when there's nothing to say except to thank you so very much, sweetheart, for the lovely three weeks' plant hunting, and to utter the cliché of all clichés — that I love you with all my heart for ever and ever, Oh *so* much, darling; if I could only tell you *how* much. Must stop now, as there's lots to do. God bless you, sweet, all my love always, from your *Jean*.

Throughout K-W's Manipur diaries, and those in the years to follow, he constantly repeats his devotion to Jean and his gratitude for her care and thoughtfulness. On 5 November 1948 he was feeling ill from neuralgia, nausea and a sense of lassitude: 'Jean looked after me like the little angel she is. She went out to collect seeds, saw the doctor, brought back some aspirin as an opiate and some lovely autumn-coloured leaves which she put in our room.'

November 6 was his sixty-third birthday: 'Darling Jean had a lovely birthday present for me. A magnificent day so we decided to go to the West Mountain after all. Jean brought brandy and aspirin in the lunch tin to succour me in case of need.' The lovely present was a carefully hoarded tin of peaches.

After visiting friends in India, K-W and Jean travelled to Manipur with Dr S. K. Mukerjee, Curator of the Indian Botanical Survey, which was sharing the expedition. He brought his wife and young son and daughter. At Imphal they were treated as honoured guests, and K-W was invited to lecture to the students at Imphal College. It was an event which came close to disaster, as he recalled in *Plant Hunter in Manipur*: 'The low dais was rather small, and in the course of my lecture, while tramping to and fro to give a visual rendering of the Great Himalayan Range at the high altitude, I stepped over the edge, to the intense joy of the audience. Luckily I kept my balance; but it was a near thing — almost a technical k.o.'

On 27 February 1948 K-W, Jean, and the Mukerjees set off in trucks to establish the expedition base at Ukhrul. K-W and Jean moved into a tiny house, little more than a shed, which they dubbed 'Cobweb Cottage (alias Bug Bungalow)'. As their collections grew they found conditions in their first home increasingly uncomfortable. In a letter to Winifred, K-W explained:

> Neither Jean nor I is very tidy, and though when we go away we leave Cobweb Cottage (or Bug Bungalow) fairly tidy, two days after our return it looks like a junk shop which has been hit by a tornado. We seem to have much more stuff to put into half as many bags, boxes, and baskets as when we arrived; the number of things which have been put aside because they might come in useful — but didn't — staggering.

Jean quickly proved to be an excellent collector. On 16 March K-W recorded: 'Jean did a smart piece of work, first of all noticing some sealing-wax red seeds on the ground in a deep *nalla* [gulley] and then finding the fruit. I recognised it for *Pittisporum floribundum*, and found the tree. She also picked up a flowering twig of a fine laurel I could not reach.' On 19 March he noted with satisfaction: 'In the three weeks we have been here we have collected ninety-five species of plants, perhaps three hundred specimens. About a quarter of these are fit for horticulture, or perhaps one third.'

On 5 April they left for the real target of the expedition, Sirhoi, the mountain where he had found and collected the 'nomocharis' but in reality lily of which he had had such high hopes in 1946. Reports of flowers from his first collection had been disappointing, but K-W remained confident that the lily would prove to be a good garden plant. To really know its worth it was vital to see it flowering in the wild. If it was as good as his instinct told him it was, it would be the first new lily to be introduced into cultivation from the wild for many years.

On 10 April they re-discovered the lily. The first plant they found had five seed capsules that had survived the winter. Then

Jean came on one with six buds, and two more with seven buds each. They dug them up in balls of soil and carried them back to their camp. They had to wait until mid-May to see it flowering in great drifts on the slopes of Sirhoi at between 7,500 and 8,000 feet. It was remarkably varied in habit, from dwarf plants of only six to ten inches high, to five-foot giants. Their size and vigour depended on whether they were growing in exposed or sheltered positions, which suggested that the plant was very adaptable. Describing it growing in the wild, K-W wrote:'the buds are almost carmine, but when they open the inside takes on a pale blush-pink, the outside (which has a radiant satiny sheen) being rose-purple'. His confidence was proved to be justified, and the Manipur lily would indeed become a popular garden plant with the name of *Lilium mackliniae*, chosen as a tribute to Jean.

The happiest and most rewarding period of the 1948 expedition for K-W and Jean was spent on Sirhoi, which he described in *Plant Hunter in Manipur*:

> It was Sirhoi's grass slope which more than anything else celebrated the summer months, when at least a hundred species of herbaceous plants flower. The most lovely herald of summer on Sirhoi was the pink-flowered *Lilium mackliniae*, which opened its first blooms in mid-May, its last in mid-July. In June hundreds of its delicate bells trembled in the grassy slopes. The dwarf iris (*I. kumaonensis*) flowered about the same time, but over a shorter period.

There were also orchids, phajus, phamaenopsis, spiranthes, indigofera, with its delicate foliage and clusters of pealike flowers; desmodium, uraraia, hedychium, cherries, michelia, schima, corylopsis, magnolia, vitis, rhus, sorbus, ilex and photinia.

While botanically it was a glorious place, it was also infested with snakes, and despite spending a lifetime in the wilds of Asia, K-W never lost his terror of them. On a collecting trip on the mountain on 20 July he stepped on one:

> My foot came squarely down on him before I saw what I was doing – it was no use pretending I had not trodden

on him. In one swift look I saw a coiled and darkish snake, which I took to be between two and three feet long, and was certainly fairly fat, not less than two or three inches around the middle. Luckily I could not see its head, which was tucked under a coil – there were not less than three coils and my foot covered their diameter. I leapt backwards on to a rock and immediately made for the path, which was close. I hailed Jean who was ahead, and stood there panting and feeling quite faint for a few minutes. Dear Jean, who came running back to see what the matter was, thinking I had hurt myself, said afterwards she had never seen me so white. I had had a real fright; and the funny thing is I had had a premonition all the time we were out that I should see a snake, and being frightened of them, I had walked carefully in the grass, watching my step – except one.

There was another moment of excitement when they heard of a red, black-spotted lily which grew in the paddy fields. It sounded suspiciously like *Lilium davidii*, but a local schoolmaster insisted that it was a native plant which was considered a great rarity. In the end it turned out to be *Lilium henryi*, which had almost certainly been introduced from China to be grown for its bulbs, which were used for food.

Throughout June and July the rains reduced the amount of collecting that could be done, and in 1948 they were particularly heavy. K-W described them vividly in his journal: 'There is a cave-like dimness as the mist creeps up like the distant sea; the drip, drip of the rain from the moss might be the ebb-tide leaving the sea-wreck bare. Spectres of wrecked ships with broken masts and battered hulls and tangled spars peer through the greenish half-light.'

The endless rain and the all-pervading damp and enforced idleness induced fits of depression, such as he described in *Plant Hunter in Manipur*: 'Occasionally into the life of the plant hunter there comes without warning a day of utter depression, a blank, empty, grey day when one does not know what to do, and does not want to do it anyway. One feels physically slack and mentally despondent. Suddenly, for no apparent reason, life has become a vacuum, and one is in the midst of it.' He did not suppose this was peculiar to plant hunters, 'but the life of the plant hunter, with its weeks of restless boredom illuminated by flashes of ecstasy, is peculiarly exposed to such attacks of indolence, the result, perhaps, of alternating tensions'.

Whatever tensions there were they did not intrude on K-W and Jean's marriage, even when conversation gave way to long silences. She explained: 'We were both people who could do with plenty of silence. I was not a chatterbox. Frank and I got on very well in that respect because we didn't have to be chattering all the time. But it was very companionable.'

A great deal of their time during the monsoon was spent in the tediously repetitive job of keeping the dried herbarium specimens free from the mildew which covered everything. Again and again they had to be taken out of their papers and brushed clean. On 7 August 1948, K-W recorded:

> This is the time when one is apt to reach the lowest depths of depression. And then suddenly the rains draw to their conclusion, and the sun shines, the atmosphere begins to dry up, the trees, many of them, begin to turn brilliant red, yellow, orange, the mud disappears, the birds – different birds – sing morning and evening, gorgeous sunsets and splendid views appear behind the white curtain which has lifted at last. That time is near, only a few weeks distant. One must work steadily on while one has the chance, and the time will fly; when the fine weather comes one will want to be off immediately, a new surge of energy will take the place of one's present lethargy, and there will not be time to do all that one wants to do.

In October they were busy with the seed harvest on Sirhoi. They also dug up nearly three hundred *Lilium mackliniae* bulbs with ice picks, and selected about two hundred of the best. These they packed in moss in bamboo baskets, which they took with them to Calcutta. Once there, they re-packed them, still wrapped in the original moss, and air-freighted them to London in January 1949.

It is appropriate here to complete the story of the introduction of the lily into cultivation in England. The bulbs arrived in perfect condition, plump and ready to make roots. They were sent to Colonel F. C. Stern (later Sir Frederick Stern) at Highdown, near Worthing in West Sussex, one of the greatest experts on lilies of this century. Eighteen months later he exhibited a superbly grown plant in full bloom at the Chelsea Flower Show, and 'like a debutante at a coming-out ball, *Lilium mackliniae* received the coveted Award of Merit.' The lily was also grown with great success at Ingwersen's nursery, where Will Ingwersen wrote, 'From our experience with it I consider that there is every prospect of it becoming a first-rate garden lily.'

Writing of the vindication of his faith in the lily, K-W had this to say in *Plant Hunter in Manipur*: 'The acceptance of a plant into the Temple of Flora, backed by authority for the new name, by valid description (in Latin), and inclusion in the Kew Index, is one thing; its coming out like a *jeune fille* into the great horticultural world is quite another. It is the difference between being born into Debrett, and getting there.'

Lilium mackliniae alone would have made the expedition a success, but they came out of Manipur with forty bundles of dried herbarium specimens — Jean irreverently called them hay — tins packed with seeds, baskets of orchids and boxes of bulbs. For K-W it had been a good expedition. In spite of the postwar austerity he had been able to re-establish himself as a professional plant hunter, and was now married to a woman who wanted to share his job with him. For the first time his expeditions were not only the pursuit of his profession; they had become an enjoyable experience that he could share with his wife. He had not been able to do this with Florinda, and the nearest he got to this happy state was in his long letters to his sister, Winifred. With Jean he had someone with whom he could experience each triumph and despair as it happened.

They enjoyed each other's company, and however slender their resources they always found the means to mark a special occasion. On 12 November 1948, when their stores were running low, they celebrated their first wedding anniversary in style. At teatime they had, according to K-W, a cake 'as tall as the Empire State Building, and dear Jean planted one candle on the top and lit it'. After a drink before dinner, there was roast duck, followed by tinned peaches and tinned Christmas pudding. 'Altogether a day to remember.'

The only shadow that fell across this unusual honeymoon — for if a honeymoon is the start to married life, then that was what the Manipur Expedition was — was the apparent indifference of Dr Robbins, the Director of the New York Botanic Garden, whose employers had put up most of the money for the trip. K-W had written regularly to Robbins reporting on progress and sending seeds, but received no replies. On 26 November he noted in his journal: 'If Robbins can't write, surely one of his staff can at least acknowledge our letters and seed packets. I am getting fed up with this lack of elementary courtesy.' In fact he did get a letter from Robbins the following day, but it did not improve his mood: 'We got an unexpected mail in the evening — a very cool letter from Robbins at last. He is "satisfied" with our results, though only the future can decide what we have accomplished — if anything.'

By 6 December they were behind schedule for their departure, and were running out of supplies, so they returned to Tocklai in Assam, which they had decided upon as a base at which to put their collections into order. From the peace and quiet of the Assamese tea gardens they looked back with justifiable satisfaction upon a successful expedition, and, to their surprise, their view was reinforced by the undemonstrative Dr Robbins, who asked them to organise a second plant hunt which the New York Botanic Garden would back with $8,000. For the first time since the end of the war K-W felt financially secure, and able at last to repay the debt he felt he owed Winifred: she had started a school for children and young people with speech defects, and had trained and employed Pleione as a speech therapist, and thus launched her career.

At Tocklai, K-W and Jean enjoyed a comfortable life while they put the products of the Manipur Expedition into order. They had a suite of rooms in a comfortable bungalow, servants, regular meals, and bridge and dancing at the nearby Jorhat Club. There was a brief trip to Calcutta to equip the second New York Botanic Garden expedition, and with it a hectic if brief bout of social life. At one party K-W was introduced to two boys who had just left Haileybury. They had asked to be introduced to him, but if they were seeking encouragement to become explorers they certainly did not get it. K-W '... talked to them in an avuncular way, showed a somewhat spurious interest in their affairs and doings, and bade them goodbye. However, they looked nice wholesome-minded English schoolboys, though I couldn't help feeling there wasn't much romance about them, or any visible "pioneer" spirit. Still you never know. Forty years on, one or other of them may be the man on the moon.'

By mid-March 1949 they were in the Mishmi Hills for what K-W called the Kamlang Expedition, and things were not going well. His camera was damaged in an accident, and three dozen plates were ruined. 'This means no flower photographs this year,' he recorded, adding miserably: 'My photography always seems to go wrong. Altogether a bad start and I do not feel optimistic about the Kamlang Valley Expedition.'

It was a difficult trip, aggravated by a plague of petty theft. One of the worst offenders was a hymn-singing servant called Jehishel. When he was caught trying to steal tea, K-W noted scathingly: 'All rather amateur and petty-theft like, typical of a bumfaced evangelical.'

Despite a bad start, 1949 was a busy year with three separate forays into the Mishmi Hills, the Khasia-Jaintia Hills and the Naga Hills. In August there was bad news from America. A letter from the New York Botanic Garden informed them that they would not be engaged for a third year of collecting. At the same time a projected American lecture tour fell through, and thus they again faced the likelihood of unemployment.

Moreover, the injury that K-W had sustained in India during the war was manifesting itself in spinal arthritis, which did not respond to treatment, and on top of that his doctor told him he was suffering from anaemia and high blood pressure.

They did consider returning to England, but instead decided

to use their own resources, and what support they could get from private backers, to organise a new expedition. The problem was where to go. The relatively free-wheeling days of the Empire were over. Now in Communist hands, China was impossible: Tibet, threatened by China, was hardly better; Korea was on the brink of war, and while India, Assam and Burma were not closed, independence had brought about new difficulties, mostly created by bureaucracy.

The changes wrought by the war in the peaceful wildernesses he adored depressed K-W. In 1949 he wrote in his journal:

> However much one may denounce material civilisation — and some aspects of it are deplorable — the fact remains that no races which come in contact with it can resist it. These hill tribes of Eastern India who are so proud of their way of life, so independent, and so contemptuous of the plainsman, have not proved immune to the subtle influences of material prosperity since the war. The Manipur Road shows the most deplorable ribbon development, shacks of all kinds, shapes and sizes, built on the verge of the wide road, while the villages are roofed with nasty corrugated iron.
>
> A fair number of wheezy lorries are Naga-owned and carry loads of potatoes to Dimapur for shipment to Calcutta. Here and there a derelict lorry stands by the roadside, half stripped and rusting slowly. There is amongst a dominant minority a hankering for money, and the material comforts that money brings. This is due partly to Western education, and partly to direct contact; whatever the cause, it is real, and must have a profound influence on the village economy.

Later he wrote of post-war Sadiya having been transformed into 'anything but an Assamese town ... a dreadful looking bunglotrocity'.

K-W and Jean decided to hunt for plants in the Lohit Valley, close to the border between Assam and his beloved Tibet. One of their backers was the New Zealand Rhododendron Society,

Of Tea, and 'Planes, and Lilies

which bought a £170 seed share. At the start of the expedition in early 1950 K-W wrote in his journal:

> Sometimes I ask myself, is it right to spend so much time, energy and money, not to mention thought, and wear and tear, and effort, to achieve what appears to be so little? Say half a dozen first-class plants. But why not. How else would one spend the time? These plants may give pleasure to thousands for many years. It is a justifiable way of spending one's life.

10

Earthquake

THE 1950 expedition got off to a good start. Jean had learned a great deal since Manipur, and now took on the role of quartermaster with enthusiasm and efficiency. Although he was a brilliant collector and explorer, K-W was an impractical man when it came to equipping an expedition. As far as he was concerned all that was needed was his scientific equipment, tents, basic stores, iodine and quinine. For day-to-day food he relied largely on local sources. This could be precarious, and had led on previous occasions to clashes with his travelling companions. It was one of the reasons why he could never have made a success of a large expedition. Jean on the other hand insisted on much better rations, good medical supplies, and even some creature comforts.

In March 1950 they crossed the border into Tibet without any problems. It was like the old days. However, K-W was already uneasy about the future of the country, as his journal entry of 23 February made clear:

> Most of the leading Tibetans are pro-British, and curb the power of the monkhood very considerably. At present little can be done because the Lamas have a majority in the Assembly — 32 out of 50, or thereabouts. The high Tibetans know that the Lamas are useless parasites, and imposters. All the Tibetans are anti-Communist, and prob-

ably in Tibet the greatest barrier against Communism could be built.

There seems little chance of 'Chinese Tibet' as far south as Atuntzi or Muli going Communist unless considerable Chinese forces enter the country, in which case they would have to fight.

In October 1950 China's People's Liberation Army marched into Tibet.

K-W's back was still troubling him, and his mood was not improved one night when, while they were entertaining an American missionary couple, their adopted Tibetan baby 'proceeded to spend a huge stinking penny in the very middle of our tent in the middle of dinner!! I think we took it rather well without a murmur.'

In April K-W and Jean were collecting in the gorge of the river Di Chu, and encountering dramatic cliff climbs. Throughout his long career K-W had never overcome his fear of heights, and now in his mid-sixties he found traversing cliff faces more terrifying than ever. One gully crossing had to be made on a nine-inch wide bridge made from two pine logs. 'I funked it,' he recorded. Eventually he did make the crossing, but not before a third log had been laid. In one day they completed a seven-hour march and collected twenty species of plants, which comprised more than one hundred specimens, all of which had to be put into presses before they could turn in for the night. The strain led to shortness of temper, and one entry in his journal reads:

> I blamed her [Jean] for climbing the big rock and collecting *Rhododendron bullatum* before I had seen it growing. This almost completely spoiled her triumph, especially as she had taken a risk to get it. I sometimes get horribly selfish and self-centred where plant hunting is concerned, as for example streaking ahead along these difficult tracks, eager to get into the big stuff, leaving poor little Jean struggling in the rear. She was terribly upset for a few minutes, but all was forgotten in our mutual triumph.

The position was made worse when his coolies refused to set up camp above eight thousand feet. They needed to be between

Earthquake

nine and ten thousand feet to reach the best of the rhododendrons. K-W wished to camp right among the alpine flowers so as to avoid long daily climbs which would have been too much for him at his age. Stores began to run short and they now had to rely on the barely adequate local supplies. While they did not go hungry, their diet lacked vitamins and minerals, and they gradually became under-nourished. On 11 June a pony carrying two months' supply of rice fell into a river and was killed. The rice was lost.

Plans for a serious foray into Tibet were now scuppered by a letter from Lhasa politely refusing them permission to enter the country, because of the threat from China. In a tone of some desperation, K-W wrote in his journal: 'It is of the greatest importance during this crucial month that we do not let despair, frustration, boredom, heat, monotony of food, or anything else get us down. Jean is splendid, simply wonderful. Poor sweet, she is even more disappointed than I am, but her morale keeps high.' He kept himself going by taking long walks and botanising as much as possible. He noted in his journal that he found himself having rather ridiculous conversations with imaginary people as he tramped along.

June passed and most of July, and they were still unable to make much progress, having been badly let down by the local man who was supposed to be hiring coolies so that they could move on to Rima. 'One and all the local people have let us down, exploited us, broken promise after promise. Gratitude for favours shown is unknown. Anyhow, now we know, I shall not give this fellow any more money, or presents,' he wrote angrily. Eventually they did get porters and were able to go to Rima in early August.

On the night of 15 August disaster struck in the form of the great Assam earthquake. It was one of the biggest ever recorded (about 6.7 on the Richter Scale) and K-W and Jean were camped close to its epicentre. There are many written accounts of the earthquake, but none is quite so vivid as K-W's description in his journal, which he began within hours of the first devastating shock:

> At first a slight tremor about ten our time, roughly half an hour after dark, followed almost immediately by a terrible

'quake, accompanied by an awful din; it lasted fully three minutes, easing very gradually, while we lay flat on our faces outside the tent, sick with terror, the earth heaving and bucking, and the noise becoming ever louder.

Now all was noise, confusion and sickening movement. Jean and I lay side by side close together flat on the ground, holding hands. Perhaps a full minute had elapsed since the first incredible movement when our world began to reel. I was dimly aware of the disaster which was happening all round us. The ground beneath us shook and rocked and heaved until I felt certain that an enormous crack must open immediately and engulf us all. But for the most part it was as though a series of irresistible blows delivered by a steam hammer were beating against the foundation of the world with the rapidity and cruel persistence of a kettle drum.

How long had we to live? How long before the thin surface ground crust crumpled beneath us, and let us fall down, down headlong into the steaming heaving bowels of the earth? Before the steam hammer dealt such a blow that our hearts would cease to beat?

The din was awful. Besides the uproar of the earthquake itself, a noise impossible to describe because it was like nothing I had ever heard, the very mountains themselves seemed to be tumbling into the narrow valley.

The crack and crash of rocks on every side was appalling. Sick with terror, though I had now so far command of myself and I was able to speak calmly to Jean who answered as calmly. I wondered how long we had to live before being engulfed in the vast rent, or buried alive, yet an inner voice told me quite clearly that it was not the end for us.

Their two servants were camped close by and K-W and Jean could see their shadows against the canvas of their tent. One of the boys, Arke, was only fifteen years old; the other, Phak Tsering, somewhat older.

Earthquake

K-W wrote two versions of what happened. In a later account he had Jean holding Arke's hand, but in the journal entries written only hours after the main convulsion, he described the scene thus:

> Jean and I lay side by side very close holding hands, then I put my arm round her. We yelled to the boys to come out into the open. They joined us, stumbling and falling, crawling, running, lay down. A [Arke] was, I think praying [in the later account he was gabbling Tibetan prayers at a great rate], PT [Phak Tsering] was crying. Jean and I spoke comforting words to each other and to them. I held A's hand.

The dark was intense, they could hear rocks cascading down the mountainside, but the immediate danger soon appeared to have passed, and they were able to return to their tents. It says a good deal for K-W's calm that he could sleep on and off during the remainder of the night. Jean could not. In one wakeful period he got up and wrote:

> It is now two hours later. Tremors are becoming fewer, and slighter, at longer intervals, but still occur. The thunder of the rock avalanches is still loud and long. Immediately after the big shock had ended there were a number of sharp explosions in the sky, like bombs exploding or gunfire.
> After the 'quake the sky became quite black, filled with fine dust. The stream with a fine head of water from which we draw our supply, has become a mere trickle. A wind, not strong, from the south got up. Barometer steady, altitude stayed at 5,000 feet, temperature 73 degrees at 10.30, a degree or two higher at the time of the shock perhaps.

At around midnight it began to rain mud because of the amount of dust in the atmosphere. The following day, 16 August, K-W measured the altitude and found it had risen by an astonishing two hundred feet.

> We looked out on a new landscape, dimly seen in the dusk-dealing dust, which was very thick to north and south. Almost every one of the dozen huts in Rima was unroofed, every annexe tumbled down.
> The Gompa at Shiga [Shigathang] was a ruin, lying half on its side; so was the Chorten beyond it. The square timber houses built of solid logs or solid planks were intact; even some of the grass hovels had been thrown down and every Mani pyramid. Deep cracks and fissures crossed the dry fields. They were very close to one another sometimes.
> The Lohit and the La Ti both seemed to be composed of liquid mud, dark coffee-coloured, though the La Ti is so covered with light-coloured foam as to appear different, except where the water pours over the boulders.

The bridge spanning the Lohit had been also destroyed.

> It was impossible to see the mountains properly until the dust, which is thicker than the March fires' smoke pall, had drifted away, or been washed out of the sky. The hills have had strips a half mile long torn off them. Wide belts have been peeled away, deep wounds opened in them from base to summit.

Surprisingly, life returned to normal with extraordinary speed, and during the morning following the earthquake, K-W heard a small boy scaring the birds in the paddy fields, and watched swallowtail butterflies foraging for nectar among the brilliant blue flowers of the cerastigma. Nobody was killed or injured, and most of the domestic animals – pigs, cattle, ponies and poultry – escaped unscathed. But water soon became a serious problem. 'All the water channels have ceased to flow, several grinding mills are in ruins. Everything is covered by impalpable dust. Lack of clean water is rather serious,' K-W recorded. A man arrived from Kahao with news that it had been flattened and had been evacuated, and a Mishmi village on the La Ti had been destroyed.

On 17 August the altitude measured out at 4,980 feet. Rock falls were constant, and tremors continued at regular intervals. The

Earthquake

paddy fields had drained, which spelled possible starvation for the hill people.'The up-valley wind brings dense clouds of white, more probably black, dust and by midday one can hardly see across the Rima basin.'

The day following the main 'quake, Jean fell ill with fever. She was in a state of shock, and later concluded that she was suffering from a nervous breakdown. It had been a difficult expedition and this was the final straw. Her condition was not helped by the constant earth tremors. On 18 August, K-W wrote in his journal:

> Poor little Jean still cannot sleep – she who used to sleep for eight hours so easily. It can only be shock. She is very brave about it, but I am terribly worried. Tremors in the night and again at about six this morning – very mild and far between; reassuring. The only thing to do is to get out how and when we can – a sad ending to our high hopes, long sustained in the teeth of frustration, and apparently on the eve of fulfilment.
>
> After breakfast I gave Jean a small shot of morphia, hoping she would sleep. She became very drowsy, but did not sleep, and after two hours vomited; this continued at intervals all day, and even when I gave her brandy with a cup of tea she brought it all up again. I am desperately worried about her. This lack of sleep is appalling, evidently due to shock. As for vomiting, she says anaesthetics always have that effect on her, so that may not be so bad as it looks. She *must* have sleep and food soon – until today she has taken her food pretty well, unnutritious as it is. She has only half a degree of fever now. Poor darling Jean – it is all my fault. God knows what will become of us if the road to Walong is badly blocked ... All my efforts must now be directed at getting Jean and the two boys out of this mess; there is no question of any more plant hunting.

The fine dust was becoming intolerable. It got into everything — food, eyes, mouth and lungs. Their clothes were caked and filthy with dust and sweat, and there was no clean water to wash them in.

Twelve hours after the morphine injection Jean was still unable to keep any food down, not even a sip of water, but by the evening she was a little better, although very weak. 'She's gone through too much this year. I will get her down the valley somehow. The future may look dark and uncertain, but the clouds will roll away presently. After dark the sky was quite overcast except for a single large hole over the western sky, where several bright stars shone. I took that as a sign of help and better days to come.'

In the back of the journal for 1950 is an entry in Jean's handwriting. She clearly felt that death was near, and it comprises instructions about keys, stores lists, and her property in store, including 'your letters to me, darling Frank — every one you ever wrote to me — in two packets. These have been my joy for years. Keep them, darling, they are yours and mine, part of our life together.'

Following her Will, she writes what is really a last love letter:

> I wish I had more to leave you, my darling, but my love for you seems to be more than my material goods. Thank you, thank you my beloved for all the sweetness you have given me. No woman has ever been happier than I, or ever can be. God forgive me if I ever let you down in my life. Trust in God's mercy, Frank, that He will surely let us be together through all eternity. He will not try you more than you can bear; He loves you and me, and all men. Have firm faith in His goodness.
>
> The peace of God which passeth understanding keep your heart and soul; and the blessings of God Almighty the Father, the Son, and the Holy Ghost be with you always through Jesus Christ Our Lord who loves you and died for you. With all my deepest love, your adoring wife, Jean.

Earthquake

Apart from worry about Jean, K-W's feelings of frustration and disappointment were acute. They had now been marooned at the bottom of the Lohit Valley for four months for lack of coolies. However, only a few hours before the earthquake struck he had written a letter to Winifred in which he claimed all their worries and difficulties were at an end. 'All is fixed for a start into the alps tomorrow and in 3 days we are promised an alpine camp at 10,000 feet or so, after nearly six months directed to that end.'

By 21 August Jean was beginning to recover, but K-W, affected by reaction, became bad-tempered and irritable. He was also plagued by a nagging pain in his back, and a stomach upset. The experience of the earthquake had had a deep effect on Jean and her outlook upon life. As an agnostic with no commitment to any particular religion or faith, K-W was fascinated by what happened:

> She has turned back completely to [the] Christian faith, and finds strength and healing there. This after fifteen years of doubts and questioning and lack of faith, she says. Of course she has always been animated by Christian principles, but now she says her prayers, has no more fear, turns to Jesus as a very present help in trouble, openly acknowledges her complete faith in him. She is terribly sincere, and it will last all her life and guide every action.
>
> She is calm and collected, talks lovingly of her parents, who she so longs to see to tell them how much she loves them and what they mean to her.

A few days later he wrote: 'Jean slept for five hours. I think she has got over her secret terror, conquered it, driven it out. She had the Testament in her hand all night. Seems much brighter this morning. Almost normal again.'

In fact the terror remained with her for a very long time. When they reached safety she could never remain in a building until she had made sure of an escape route. When she visited a friend who lived in a block of flats in the Cromwell Road in London, close to the underground railway, she went ashen when she felt the vibration from a passing train. K-W, on the other hand, recovered quickly from the effects of the earthquake, but he was tormented by a sense of guilt, blaming himself for exposing Jean to such danger. 'All this has come upon me because of the lust of ambition – to succeed where others have failed, and in the teeth of all frustration, where success seemed unlikely.' By September the position was becoming serious. They had had to give rice to a patrol of soldiers whose camp and stores had been buried in a massive rock fall, and now their reserves were very low. K-W's original plan to stay quietly in their camp until the river fell and the bridge over the Lohit was repaired now seemed impracticable. There was no rice to be bought, and they had rations for only about ten days. Six days later there was a large rock fall which sent white dust pluming into the air - 'the spinning, bouncing rocks enlarged by the dust to terrifying size hum through the air'. Despite all the difficulties, K-W did manage to get out again onto the mountain and collect what seed he could find, including that of a fine yellow cornus.

Shortly afterwards they were able to leave for Walong, and had not gone far before they met a party of men of the Assam Rifles who had been sent to rescue them. Jean was still weak and had to be carried for the first part of the trek. Their route took them through scenes of great desolation, with thousands of trees shaken out of the soil. There were rock slips all around them as though the whole country was on the move. K-W noted: 'A rumble of a heavy fall gave way to a crashing, grinding roar, as the fall gains way with terrific explosions as big rocks spinning through the air hit the boulders below. Clouds of white and yellow dust arose, and through the fog one saw stones whizzing, bouncing, ricocheting into the river.'

When they reached safety he wrote to Winifred that the journey

> almost turned my white hair black, or turned me as bald as a billiard ball.
> Not a day but we carried our lives in our feet, crossing precipices, clinging like lizards, balancing, moving step by step, afraid to look down a thousand feet to the roaring

Earthquake

river, equally afraid to look up to where cascades of boulders might at any moment bombard us, if the cliff loosened up. At times it was a nightmare. Jean, who has a steady head, did not mind this bit, while every cliff traverse (one was nearly half a mile) made me sick with terror.

In the same letter he wrote that a year's work was practically ruined: '... rather a disappointment for what is probably my last plant hunt'.

On 30 September he recorded: 'Our plant hunt is miserable, less than 750 species, and less than 50 species of seeds (unless we include some tropical species). We have done our best, but fate has been too much for us ...' He felt so keenly that he had failed and let down his backers that he wrote to Dr J. S. Yeates of the New Zealand Rhododendron Society, asking how much of the society's investment in the expedition he should return in view of the small quantity of seed he had been able to send.

Dr Yeates replied generously: 'First let me say that we are all very relieved to hear you are safe and sound. So far as the contribution through our Association is concerned ... my personal feeling is that it is our bad luck as well as yours, and that you presumably incurred the expenditure and put forth the effort expected of you.'

On 9 October K-W left Jean at Walong – she still had not fully recovered – and set out for a camp at 7,500 feet. Although it was well short of the alpine belt which started at about twelve thousand feet, he believed he would be able to collect some worthwhile seeds. It was the first time he and Jean had been apart since they were married three years before, and he missed her sorely, but he was determined, as he put it, to 'drag the fringe of success out of the slough of failure'. On the 10th he climbed to ten thousand feet and got in amongst 'the big stuff' – rhododendrons, sorbus, *Rosa moyesii* and *R. sericea*. Two days later he was back with Jean, who was feeling very much better, and told him they were leaving for Sadiya on the 15th.

By 24 October the end of the expedition was in sight, and he was able to write in his journal: 'I shall be glad to get back. The old daily round, the common task won't come amiss. In ten days or so we shall be enjoying not luxury, but comfort; in three weeks or a month we shall be back in the old, unexciting routine of the Western life; in three months we shall be sighing for the mountains again. But for the present we have had enough.'

On 4 November they were in Tayju and he felt able to write: 'God has been very good to us. I do feel a divine providence has been looking after us ...' On his sixty-fifth birthday, 6 November, they arrived in Sadiya.

11

The Last Lily

AFTER a fairly long rest at Shillong, K-W and Jean returned to Tocklai to work on the plants and seeds they had been able to collect in 1950, and to prepare to return to England, and an uncertain future. They were not sure where they would live, and had come up with a scheme whereby they, Winifred, Martha and Jean's parents should pool their resources and buy a communal home – the patriarchal system, as Jean described it. It remained an idea.

For most of 1951 and 1952 they were in England busy preparing for a new expedition. Because of the many problems attached to returning to the Himalaya, at least to the Sino-Himalaya, this time they seriously considered an expedition to Papua New Guinea. 'It might have something to offer, and at least it is not yet infected with Communists,' K-W remarked at the time. Despite these thoughts, Burma remained their first choice, and they planned a collecting trip to North Burma that would cost £2,300. Part of this money was promised by the Royal Horticultural Society and the Natural History Museum. Members of the New Zealand Rhododendron Society bought shares, and a commission to collect seeds of tender rhododendrons for the important New Zealand nurserymen, Duncan and Davis, also helped towards raising the finance. They were eventually given permission to collect in North Burma, and on K-W's sixty-seventh birthday, 6 November 1952, they left London for Liverpool, where they boarded the Bibby Line steamer *Staffordshire*, and arrived in Rangoon on 11 December. Their permit to travel was issued on the condition that they agreed to make a collection of plants and seeds for the Burmese Government, and at the same time train two Burmese men as collectors and field botanists.

On 7 January the party reached Sumpra Bum, and from there travelled to a village called Hkinlum, where they established their base camp. For K-W and Jean home was now a two-roomed bamboo hut in sight of a magnificent mountain, Hkangri Bum. They were in Kachin State, where K-W was well known as Nampan Duwa — the Flower Chief. In the weeks that followed the work of hunting and collecting went ahead well, but it was not until 2 April that K-W made a discovery that marked the expedition as botanically memorable.

He had climbed to about 5,500 feet when he spotted a very large tree festooned with epiphytic plants — ferns, orchids, moss, hedychiums, *Vacerium serratum*, agapetes, gisnarada, and tucked in amongst them two lily plants in fruit. There was no mistaking the fact that they were lilies. He could clearly see the open capsules. What made the sighting unique was that there were, until then, no known epiphytic lilies. What made it maddening was that he could not reach them. He was remarkably philosophical, and jotted in his journal: 'Alas, no chance of getting it. However, there must be more.' He would not forget it, but meanwhile they had to get to a higher altitude for plants that would be hardy in temperate gardens.

They were not sorry to leave the base camp for a while. It was not a comfortable place, and they were plagued by cockroaches, flies of all kinds, and giant jungle rats which feasted on anything they could find from soap to plastic bags and boxes. Sundays at the base camp were also something that would not be missed. K-W wrote: 'Sunday is a cross we have to bear. The local Christians attend service twice and sing horribly. Between these execrable bouts of piety — American Baptist Hymns — they crowd round our hut and gaze at us till their eyes go pop.' On 14 May they were climbing to the alpine level in bitter cold and driving rain, with K-W dressed in the most unsuitable way. Indeed he

The Last Lily

rarely dressed properly for the conditions which he knew so well from experience. On this occasion he cut an odd figure in a thin cotton vest, thin tropical bush shirt, cotton pyjama trousers and shorts. Two days later however he made himself as weather-proof as possible: 'Put on a thin cotton vest followed by a woollen one, followed by a cotton shirt — all short-sleeved — I also wore khaki shorts — ragged — woollen socks — full of holes — and strong boots. Over all I wore a plastic waterproof with hood, and a rain hat inside it.'

By 2 June they were back at the base camp, and able to celebrate the Coronation of Queen Elizabeth II in fine style:

> First sat out on the veranda and ate peanuts, pickled walnuts and maraschino cherries, and then to dinner, roast cock with fried potato balls, boiled pumpkin, soup, followed by Christmas pudding. We drank the Queen's health in brandy, and went out to light our sparklers — which proved somewhat uncombustible to start with. However, it was fun, all the servants joined in so that we had a number sparkling at the same time.

Five days later he made another important discovery — a fallen tree covered with epiphytes, including several lily-like plants with fairly broad leaves. He was immediately reminded of the inaccessible lilies he had seen in April, and which he called the Arahku Lily, because he first saw it at a spot within sight of the Arahku peaks. He dug up two bulbs and took them back to a little garden he had made near their bamboo hut where he stored living specimens of plants they had collected. He planted one in a tree in full light and one on a tree stump in the deep shade of what he called the Orchid Wood. The only disappointment was that the plants did not appear to be producing any buds. From the capsules he had seen he guessed they would bear fairly large flowers, not as large as *Lilium auratum* or *L. regale* but probably the size of *L. nepalense* or *davidii*, and that they would grow from two to three feet tall. Altogether the find was exciting. No new lilies had been introduced from the wild since he discovered *L. mackliniae* in Manipur, and for a man who knew his career was drawing to a close, to collect another hitherto unknown lily was something of a coup.

'Should it turn out to be a purely epiphytic species it will surely be unique; there is of course the unpleasant possibility that it is neither a lily or a nomocharis, but a new genus altogether – unpleasant because horticulturists will fight shy of an unfamiliar name.'

Shortly after finding the lily they were on the move again, and on 16 June reached a camp set up for their attempt to reach the summit of a mountain called Tama Bum. It was to prove a horrible experience. K-W described one day's climb thus:

> I found the first two hours very exhausting. By the time Jean caught me up I was going badly. We went on for about an hour climbing steeply, and at last steadily, without any going down. By eleven o'clock I felt so whacked I really didn't know how to go on – a horrible feeling. I felt as though I might drop dead at any moment – and almost wished I could. But darling Jean was with me cheering me on and I knew I had to make it.

On 25 June he had a brief blackout, but two days later rallied because Jean had fallen ill and was confined to bed with a heavy cold in their sodden bivouac. He wrote on the 27th:

> Jean is not seriously ill, but I feel she is slowly deteriorating, in spite of courage and determination. She has a great reserve of staying power, but it must not be drawn on indefinitely. The expedition may have to be abandoned: and yet I don't think that either of us is capable of doing the journey back to Sumpra Bum during the monsoon. I feel dreadfully depressed: it would break my heart to have to give up now, but if anything should happen to Jean I should die. I can hardly bear to see her lying in the tent day after day, with the mist swirling round, the rain falling steadily, the forest gloomy and cheerless. I should have known how it would be and insisted on returning to SB at the end of May at the latest. And yet at the bottom of my

The Last Lily

heart I still think the expedition is going to turn out a success.

The monsoon, which had now settled in, invariably brought on fits of deep depression. On 6 July he was writing:

> Jean is too exhausted to sleep. I felt too ill. A cock finally woke us at four. By four-thirty we felt thoroughly ill, dispirited, whacked to the limit. As close to a final defeat as makes no difference. I had frightful neuralgia, could only whisper, with a hacking cough. Jean was worried beyond endurance. We must return to England at once or we shall slowly die where we were [sic]. Or at least we must get back to Sumpra Bum where there was a doctor. But shall we ever get there in our present state? And we were not fit to start at all really, nor strong enough to stand the jungle marches. We must stay and hope for the best. Perhaps in a few days we shall feel better.

Just how oppressive this season was can be gleaned from his description of the rain-saturated forest:

> After weeks of darkness and rain, the fecund earth breaks out into noisome blisters, and fungoid pustules, throwing up saprophytic ghouls and loathsome excrescences. It becomes a shambles of low, boring, merciless low life, the stench both of the living and of the slain is nauseating.
>
> Rotting leaves, bud scales, flowers long since dead, early ripening fruit, plucked from their life source by birds or squirrels, or by deadly hordes of insects before their time, to become prey of yet other hungry hordes of worms in the festering world, throw up their vile stenches. A march through the forest in the mid-summer heat of the rains is an experience. How astonishing it is, this change from the clean air of spring, the crisp sounds of life beginning, to this dismal stinking strife and struggle for life, this smell of the grave, which after a short winter sleep, a few months hence, will give rise once more to fresh sweet scents and indescribable beauty.

It was always the thought of the next spring, the flowering season and the following seed harvest, that kept K-W going during his lowest periods; that and the extraordinary energy and optimism of the treasure-seeker which will often overcome the most dire difficulties.

Towards the end of July one of the mystery lily plants in his garden had produced two buds. They were hanging in typical *Lilium martagon* fashion, and K-W wondered what colour they would be. The lily's apparent resistance to unfolding its buds was frustrating, but around the middle of August the tips of the petals began to curl back, and when its turk's cap flowers were fully open they were revealed to be of a pale Nile green with the texture of watered silk, vermillion anthers, and a glorious fragrance of nutmegs. So that he would have some seeds to take home to England, he took pollen from one flower to another with a fine paintbrush in order to fertilise them, rather than leave it to chance hybridisation. Fortunately a woodcutter who felled a tree upon which the lily grew was able to supply them with a few more bulbs, and they found other specimens that yielded a small seed harvest. Eventually the lily flowered in England under glass at the Royal Horticultural Society's garden at Wisley in Surrey, and at the Liverpool Municipal Parks. It was named *Lilium arboricola*, but sadly did not survive in cultivation.

Another potentially outstanding plant that K-W and Jean discovered, from which they were unable to collect seed, was a tender clematis. It had chocolate-brown perianth lobes and ivory stamens. When the sun shone through the petals they turned beetroot red.

Despite a mysterious illness which laid him low in September, by October, with the seed harvest, K-W was again full of enthusiasm and optimism. On the last day of seed collecting they came across a colony of primulas that seemed to be a species that fitted in between the *candelabra* and *sikkimensis* groups, and could have been a natural hybrid. They collected a good quantity of seed and dug up several plants. He wrote in his journal:

The Last Lily

> This was a real triumph and I was elated. It seemed to me that if one had time to explore all the accessible gullies one might find quite a number of good plants here and there, small colonies or isolated plants. Since we came here a week ago we have found cotoneaster, two species of onphalogramma, primula, gentiana, crementhodium, and an aster, none of which we had seen in June, also a dwarf campylogynum rhododendron. We collected seed of nearly all our marked plants, as well as several new ones.

They were back on Tama Bum, the mountain which had taken such a toll of their health and strength, but in the end it had yielded up a botanical treasure trove. One other incident afforded K-W the greatest pleasure. While he could often be sharply critical of the Burmese botanists who accompanied them, he was deeply touched when one of them, Chit Ko Ko, carved in the bark of an aralia: 'To F. Kingdon-Ward, who knew and loved North Burma', and signed it CKK.

At the beginning of November they were collecting on their second mountain, Thulam Bum. It was 6 November, K-W's sixty-eighth birthday, and a day to remember:

> I could not have spent it better or enjoyed it more, nor had more glorious weather – and views. The results were, if not spectacular at least satisfactory and worthwhile. We collected at least five species new to the collection (dwarf ilex, notholirion, spirea, *Rhododendron megeratum*) with seeds of all except the ilex; seed of bergenia, a little more rotundifolia primula, and ample *Rhododendron martinianum*. Also seed of half a dozen species of which we badly need more.

It was a wonderfully rich area highlighted by his entries for 7 and 8 November covering two visits to a glorious natural garden in a gully, where he found:

> ... masses of a new petiolaris primula, and in the gravel spits in the bed of a stream which ran through the gully

were candelabra primulas, and the blue-black lactuca, a gaultheria with black berries, and a dwarf erect species with white berries, and scattered through it violets in fruit, and more gaultheria with lapis lazuli blue berries, and another plant, a scrophulacea, with a six-inch spike of fruit like some New Zealand veronica. And there were rhododendrons, a yellow viola, a gaultheria with scarlet, flask-shaped fruit, a new saxifrage, and masses of stellaria and epilobium.

By 6 December they were back at Sumpra Bum, and enjoying almost forgotten luxury. They had tea on the veranda at a table covered with a clean table cloth. 'In the evening sitting round the fire, we had a tot of whisky, hot peanuts and chipped potatoes, followed by chupatties, and a first-rate curry.' By the time they left Myitkyina on New Year's Day, 1954, they had covered seven hundred miles on foot, and returned with a splendid collection of plants and seeds.

While they were in north Burma they had laid a plan to go to the 12,600-foot Mount Saramati, close to the Chindwin River. It was an isolated peak which had never been explored, but aerial photographs had revealed that the top of the mountain was covered with a scrub which included rhododendrons, while the lower slopes were clad with a very varied cool temperate forest. K-W had no doubt that it would prove to be rich in new species. The expedition proved to be too expensive and too difficult to organise, and it remained ever out of his reach.

Instead, during 1955, when they were back in England, they were able to organise an expedition to the Chin Hills. Part of the backing came from a syndicate made up from a small group of Cornish enthusiasts and a group of keen gardeners in Ireland, organised by Colonel Grove Annesley, who had created a superb rhododendron garden at his home, Annes Grove, Castletownroche, in County Cork, which contains many fine Kingdon-Ward introductions.

During their stay in London, K-W was admitted to the Hospital for Tropical Diseases for a check on a persistent stomach disorder. Although all that was discovered was very high blood pressure, the specialist told Jean that if he had been asked to give K-W a clean bill of health for a Colonial Service posting he would have been unable to do so. This placed her in a terrible dilemma. Plans for their next Burma expedition were well advanced, yet was it safe for him to go? She recalled:

> I realised that he was extremely likely to have a stroke when we went to more or less high elevations, but at the same time I thought that this man has been doing this job all his life; he has no resources to allow him to sit around and contemplate, just because he has high blood pressure. I thought of George Forrest and Reginald Farrer, who both died in the field, and who was I, because I loved the man, I really loved him, to throw spanners in the works and say, 'Oh, Darling, you can't go back to Burma because your blood pressure is too high!'
>
> So we didn't talk about it at all, and as a matter of fact we got away with it. But I do clearly remember that many times, say a dozen, during that expedition in 1956 in North Burma, where our base camp was at about fifteen hundred feet and we didn't go higher than the top of Mount Victoria, which was just over ten thousand feet, I do remember then many times he had a few moments of not exactly a blackout, but looking a bit dreamy and saying his foot was tingling, which I imagine was a mini-stroke.

At the beginning of January 1956 they left for Burma, but when they arrive in Rangoon they found the country greatly changed. The place was in a turmoil, and corruption was rife. The streets were filthy and jammed with traffic, the postal service had all but broken down, and British-owned companies were closing down and getting out. K-W declared that it was little more than a police state with no parliamentary opposition.

On this expedition they were again accompanied by two Burmese botanists, plus a Swedish lady botanist who turned out to be something of a liability. Shortly after they had settled in to their base camp, she carelessly tossed a lighted cigarette aside and set fire to the mountainside. They tried to beat out the fire, but to no avail. It leapt across a narrow track and, as KW recorded in his diary:

> ... the tinder-like vegetation, dead leaves, dry grass, everything at the foot of the rocks, where a few minutes before we had been climbing, was ablaze. On the fire spread in inconsumable rapidity; grass, bushes, trees sprang ablaze. Strong winds swept it up the rocks and up the mountainside. On each side of the cliff and on the slope on the lower side of the path, the mountain was on fire. We could only watch fascinated. Presently we ourselves had to run.

Fortunately nearby villagers led by their headman were able to bring the fire under control before the damage became too great, and in the end K-W was only asked for a tiny sum by way of compensation. But he also laid on a drinking party by way of thanks. Vast quantities of *zu*, the local beer, were consumed by everyone. In a letter to Winifred he said that even the babes in arms took 'alternate doses of mother's milk and *zu*. Soon everyone was oozing bonhomie, love and eternal friendship.'

The Swedish lady committed her next *faux pas* about a fortnight later when she slipped out after dinner for an assignation with one of the Burmese botanists. When K-W discovered what had happened he was very angry. Such a thing had never occurred during any other expedition he had led. He lay awake but she did not return until after three in the morning: 'I just let her know I was awake and that I would have an explanation in the morning; I was almost too furious to speak steadily – but also greatly relieved that she was alive and well.'

The following morning he made his feelings clear to the young woman and the Burman. In his journal he observed that if she 'goes to pieces after less than a month of the ups and downs of an expedition it is clear she is not made of the right stuff for a long expedition – six months or more. She has not got the stamina, and nobody could predict her next folly.' But then he softened a little by recalling: 'How well I remember my own follies and breakdowns! How I ever survived them and managed to carry on God alone knows. That alone should prevent my being harsh with others, much younger than myself. But I think it only right to be firm.' In fact on one occasion when he was a young explorer he had managed unwittingly to get himself engaged to the daughter of a village headman, and had only extricated himself from the arrangement with difficulty.

It was not that K-W was an intolerant bigot who took a high moral tone, but he was of the old European school of Asia, which certainly did not approve of white women having affairs with the local native men. What offended him even more was that he believed that both of them had shirked their duty for the sake of a frivolous encounter.

In June Jean had to fly down to Rangoon for treatment to her teeth, which she believed were poisoning her. K-W faced the prospect of her absence with alarm. He had been having difficulties with his breathing, and his back was hurting him badly. He found life without her intolerable. He wrote in his journal: 'I am afraid I shall miss her dreadfully, and hardly know how to pass the time until she returns.'

Two days after she left he trekked to Ranchi to search out some plants needed for the collection, but on 27 June he was back at the base camp, and up early in the morning so that he could walk down the road to meet Jean, who was expected back that day. But before he could start out: '... in she walked looking radiant. She got up by moonlight and starting at the crack of dawn came tearing up the hill to arrive here by nine o'clock. It was *lovely* to see her looking so fresh and well and happy after nearly a fortnight's absence.'

Although the expedition was a success, it was not spectacularly so and by the beginning of 1957 they had left Burma and were in Ceylon for what was to be their last expedition — a gentle collecting of orchids. It was as well that this last trip was among some of the most beautiful flowers in the world, and conducted without stress or worry, being comfortably financed mainly by Swedish backers, for K-W's world was shrinking and locking him out. Burma was under pressure from China: 'violence everywhere, deliberate provocation to war ... In such a world plant hunting and peaceful gardening seem to have no place,' he grieved.

Frank Kingdon-Ward

From Ceylon K-W and Jean travelled to Sweden, where he lectured and twice met the King, a keen and knowledgeable gardener, and an expert on rhododendrons. They also made a trip to Norway.

When they returned to London K-W's mind once again turned to searching out new areas for plant hunting. He considered exploring the highlands of Vietnam, and again New Guinea. There was even a short flirtation with retirement to the Seychelles. None of these was to be.

On Easter Sunday, 1958, K-W and Jean were having a drink in a pub in Kensington. He said his right foot was tingling, and a little later that he could not feel it at all. He stood up, tried to take a pace, and staggered for a few steps before he had to sit down again. People drinking in the pub came to his aid. He had had a stroke. Twenty minutes after he was taken to a nearby hospital he lost consciousness. He was moved to the Atkinson Morley Hospital in Wimbledon, but never came out of the coma, and sixty hours later he died.

The swiftness of his death was merciful. He would have found it quite intolerable to have to endure lingering ill-health and the life of an invalid, but had he been able to choose a time and place to be felled by a stroke it is unlikely that it would have been a pub in Kensington. He would probably have preferred to end his life like his plant-hunting contemporaries George Forrest, who was felled by a massive heart attack walking on a hillside in Yunnan in China, and Reginald Farrer in his 'snuggery' high in the mountains between the Irrawaddy and the Nmai Hka.

Despite the self-doubts that so often appeared like spectres to haunt him during his life, K-W died without any cause for self-reproach, and no epitaph or eulogy can better mark his remarkable career than the superb plants he introduced into gardens throughout the Old and New Worlds.

Since he could not be buried in some remote alpine pasture in far-flung Asia, it is fitting that he now rests in Cambridgeshire, the county from which he set out on his long and magnificent adventure. He is buried in the churchyard at Grantchester, and it is almost as though Rupert Brooke was predicting K-W's last return

The Last Lily

when he wrote in his poem, 'The Old Vicarage, Grantchester':

> 'God! I will pack, and take a train,
> And get me to England once again!
> For England's the one land, I know
> Where men with Splendid Hearts may go;
> And Cambridgeshire, of all England,
> The Shire for Men who Understand;
> And of *that* district I prefer
> The lovely hamlet Grantchester.'

Expeditions[1]

1909–10:	Western China (Shanghai to Tatsienlu (Kangting) and into South Kansu).
1911:	North Yunnan and Tibet (T'eng-yueh, Tali, Atuntsi, Menkong, Dokar La, Batang, Gartok).
1913:	Yunnan and Tibet (Myitkyina, Tali, Likiang, Atuntsi, Dokar La, Paima Shan, Pitu La).
1914:	North Burma (Myitkyina, Hpimaw, Imaw Bum, Putao).
1919:	North Burma (Imaw Bum, Hpimaw).
1921:	Yunnan and Szechwan (Tali, Yungning, Muli, Bhamo).
1922:	Yunnan and Szechwan, Tibet and North Burma (Bhamo, Tali, Muli, Atuntsi, Chamutong, Taron River, Nam Tamai, Putao).
1924–25:	Eastern Himalaya and Southern Tibet (Sikkim, Gyantse Dzong, Tsangpo Gorge, Bhutan).
1926:	North Burma and Assam (Myitkyina, Putao, Seinghku River, Diphuk La, Lohit River, Sadiya).
1927–28:	*Assam, Mishmi Hills.*
1929:	*Burma and Indochina (Southern Shan States, Upper Laos).*
1930–31:	North Burma (Myitkyina, Nam Tamai, Adung River, Namni La).
1933:	Assam and Tibet (Sadiya, Rima, Rong To Chu, Shugden Gompa, Delei Valley).
1935:	Assam and Tibet (Charduar, Mago, Sanga Chöling, Lilung, Tongkyuk Dzong, Po Yigrong, Gyamda Dzong, Ganden Rapden Gompa).
1937:	North Burma and Tibet (Myitkyina, Putao, Nam Tamai, Adung River, Gamlang River, Ka Karpo Razi).
1938:	Assam, Balipara Frontier Tract.

[1] Source: STEARN, W. T. in F. KINGDON WARD: Pilgrimage for Plants, London 1960, 13–15.

Expeditions

1938–39:	North Burma (Myitkyina, Htaw, Imaw Bum, Panwa Pass, Hpare Pass).
1946:	Assam, Khasia-Jaintia Hills.
1948:	Assam, East Manipur.
1949:	Assam, Mishmi Hills, Khasia Hills, Naga Hills.
1950:	Assam and Tibet (Sadiya, Lohit Valley).
1953:	North Burma (Myitkyina, Sumpra Bum, Hkinlum).
1956:	West Central Burma, Chin Hills, Mount Victoria.
1956–57:	Ceylon.

Books by F. Kingdon-Ward

On the Road to Tibet. The Shanghai Mercury Ltd, Shanghai 1910.
The Land of the Blue Poppy. Travels of a Naturalist in Eastern Tibet. Cambridge University Press 1913. Repr. Allen Lane Penguin Books, London 1941; Minerva Press Ltd, London 1973.
In Farthest Burma. Seeley, Service and Co. London 1921.
The Mystery Rivers of Tibet. Seeley, Service and Co. 1923.
From China to Hkamti Long. Edward Arnold and Co. London 1924.
The Romance of Plant Hunting. Edward Arnold and Co. London 1924. Repr. 1933.
The Riddle of the Tsangpo Gorges. Edward Arnold and Co., London 1926. Introduction by Sir Francis Younghusband.
Rhododendrons For Everyone. The Gardeners' Chronicle Ltd, London 1926.
Plant Hunting on the Edge of the World. Victor Gollancz Ltd, London 1930. Repr. Minerva Press Ltd, London 1974.
Plant Hunting in the Wilds. Figurehead (Pioneer Series), London 1931.
The Loom of the East. Martin Hopkinson Ltd, London 1932.
A Plant Hunter in Tibet. Jonathan Cape, London 1934.
The Romance of Gardening. Jonathan Cape, London 1935.
Plant Hunter's Paradise. Jonathan Cape, London 1937.
Assam Adventure. Jonathan Cape, London 1941.
Modern Exploration. Jonathan Cape, London 1945. Repr. 1946.
About This Earth. An Introduction to the Science of Geography. Jonathan Cape, London 1946.
Commonsense Rock Gardening. Jonathan Cape, London 1948.
Burma's Icy Mountains. Jonathan Cape, London 1949.
Rhododendrons. Latimer House Ltd, London 1949.
Footsteps in Civilization. Jonathan Cape, London 1950. Beacon Press, Boston 1951.
Plant Hunter in Manipur. Jonathan Cape, London 1952.

Books by F. Kingdon-Ward

Berried Treasure. Shrubs for Autumn and Winter. Colour in your Garden.
Ward, Lock and Co. Ltd, London and Melbourne 1954.
Return to Irrawaddy. Andrew Melrose, London 1956.
Pilgrimage for Plants. George C. Harrap and Co. Ltd, London 1960.
With a biographical introduction by William T. Stearn, including a list of F. Kingdon-Ward's expeditions.

Bibliography

Acton, Harold. *Peonies and Ponies*. Repr. Oxford University Press 1983.
Bixler, Norman. *Burma: A People*. Pall Mall Press 1971.
Byron, Robert. *First Russia, Then Tibet*. Macmillan 1933, repr. Penguin Books 1985.
Cable, Mildred and Francesca French. *China: Her Life and People*. University of London Press 1946.
Chapman, Spenser. *Lhasa, The Holy City*. Chatto & Windus 1940.
Fitzgerald, C. P. *China: A Short Cultural History*. The Cresset Press 1950.
Fleming, Peter. *The Siege at Peking*. Rupert Hart-Davis 1959, repr. Oxford Paperbacks 1984.
Bayonets to Lhasa. Rupert Hart-Davis 1961, repr. Oxford Paperbacks 1985
Guibaut, André. *Tibetan Venture*. John Murray 1949.
Harrer, Heinrich. *Seven Years in Tibet*. Rupert Hart-Davis 1953, repr. Granada 1983.
Return to Tibet. Penguin Books 1985.
Hedin, Sven. *Adventures in Tibet*. Hurst & Blackett 1904.
Hosie, Lady. *A Chinese Lady*. Hodder & Stoughton 1929.
Hughes, Richard. *Foreign Devil*. Repr. Century Publishing 1984.
Kaulback, Ronald. *Tibetan Trek*. Hodder & Stoughton 1934.
Salween. Hodder & Stoughton 1938.
Kaiming Su. *Modern China: A Topical History*. New World Press 1985.
Lanning, George. *Wildlife in China*. The *National Review* Office, Shanghai 1911.
McDonnell, F. J. *The History of St Paul's School*. Chapman & Hall 1909.
Migot, Andre. *Tibetan Marches*. The Travel Book Club 1955.
Payne, Robert. *Forever China*. Dodd, Mead & Co, New York 1945.
Power, Brian. *The Ford of Heaven*. Peter Owen 1984.
Pratt, John. *China and Britain*. Collins, 1944.

Bibliography

Rattenbury, Harold B. *China, My China*. Frederick Muller 1944.
          China-Burma Vagabond. Frederick Muller 1946.
Roy, Claude. *Into China*. Sidgwick & Jackson/MacGibbon & Kee 1955.
Rodzinski, Witold. *The Walled Kingdom*. Collins Flamingo 1984.
Schurmann, Franz and Orville Schell, eds. *Republican China*. Pelican Books 1967.
Smedley, Agnes. *China Correspondent*. Pandora Books 1984.
Ward, Mrs. *Memoirs of Kenneth Martin Ward*. Privately printed 1929.
Wei, Katherine and Terry Quinn. *Second Daughter*. Harvill Press 1985.

Index

Adung Valley (Burma), 135
Africa, 146–7
Aleurites fordii (wood oil tree), 26
Alpine Garden Society, 141
American Begonia Society, 134
Andrews, R. C., 25
Androsace wardii, 30
Annes Grove, Castletown (County Cork), 200
Annesley, Colonel Grove, 200
Arahku lily, *see Lilium arboricola*
Assam, 95–6, 100, 110, 124; K-W's wartime duties in, 161; K-W works in tea plantation in, 162–5, 168; 1950 earthquake, 183–8
Asystana castinopsis, 133
Atuntsi, 47–8

Baghdad, 56
Bailey, Lt-Col F. M.: explorations, 48–9, 69; K-W names plant for, 81
Balfour, Bayley, 27–8
barberries, 88
Batang (Tibet), 34, 38
Beaumont-Nesbitt, General, 144
Bedford, Herbrand Arthur Russell, 11th Duke of, 16–18, 21, 24–5
Bees' seed firm, 27, 29, 32, 41, 45, 54, 61
begonia species, 134
Betty Rose, 103–4
Bhamo (Burma), 29
birds: K-W's interest in, 109
Blackwood's Magazine, 98, 158
Blake, William, 8
Bodley, Sir Thomas, 1
Bombay, 160
Boo (Welsh girl), 103–4
Borneo, 15

Botanical Magazine, 82
Boxer rebellion (China), 14
Brahmaputra river, 6, 45, 48–9; supposed Falls, 47, 49, 69
British Broadcasting Corporation, 43
Brooke, Rupert, 205
Brooks Carrington, R. B., 110, 113
Buddhism: rituals, 24–5
Bulley, Arthur Kilpin, 26–8, 61
Burma: K-W visits, 51, 55, 95–6, 100, 134–6; in wartime, 149–50; Japanese invade, 151, 153–6; K-W's 1952/3 collecting trip in, 193–5; 1956 Expedition to, 201–3; conditions and politics, 201; forest fire in, 201–2

Calcutta, 155
Caltha calustris, 43
Cambridge: K-W's childhood home in, 6–7; K-W attends university, 11, 43
Cambridge: School of Botany, 17
Cambridge University Press, 43
Cape, Jonathan, 68
carmine cherry, *see Prunus cerasoides rubea*
Carrusa, 133
Carter, Harold, 25
Cawdor, John Duncan Vaughan Campbell, 5th Earl of: accompanies K-W on Tsangpo expedition, 47, 69–75, 77–9, 89–91, 93–4, 110; depressions, 75–6; tooth trouble, 78, 91; discovers new rhododendron, 85; takes sheep on trip, 90
Ceylon, 203
Chiang Kai-shek, 131
Chin Hills, 200
China: conditions in, 13–14, 25; K-W on expedition in, 16–17; religion in, 23–4;

Index

China: conditions in (*cont.*)
  and Tibetan unrest, 33–4; revolutions and riots in, 34–6; railway construction in, 35; K-W returns to (1913), 44–6; authorities obstruct K-W, 50–1, 59, 132–4; declares war on Germany, 59; political changes in, 131; K-W attempts to visit (1937), 132–4; in World War II, 151–2; wartime routes to, 153; occupies Tibet, 182
Chindwin river, 119
Chinese Communist Party, 131
Chinputang Party (China), 36
Chionocharis hookeri (Himalayan forget-me-not), 120
Chit Ko Ko (Burmese botanist), 199
Churchill, Winston S., 150
Clark, Louis, 76
Cleeve Court (Streetly-on-Thames), 116–17, 128–30, 141
Clutterbuck, Elizabeth, 102
Clutterbuck, Hugh: travels and friendship with K-W, 47, 98–101, 114; financial support from, 98; death, 102
Colet, John, Dean of St Paul's, 10
Colet Court (school), 6, 9
Cousins, Captain (of Military Police), 64
Cox, E. H. M., 44
Craigavon, James Craig, 1st Viscount, 116
Cranbrook, John David Gathorne-Hardy, 4th Earl of; travels with K-W, 103–6, 108–9; poisoned by honey, 109–10
Cutting, Suydam, 100, 119, 121, 136
Cypripedium (or *Paphiopedilum*) *wardii*, 106–8

Daily Telegraph, 145
Dalai Lama, 80
Delavay, Père, 81
Doshong-la (Tibet), 87
Douglas, David, 15–16
Duncan and Davis (nurserymen, New Zealand), 193
du Pont family (USA), 139

Edinburgh: Royal Botanic Gardens, 41
Egypt, 147
environment: human effects on, 137

Exbury, 98, 142

Fairchild, David, 25–6
Farrer, Reginald,. 44, 201, 204
Fisher, Mr (scientist), 6
Forrest, George, 27, 34; death, 201, 204
Fort Hertz, 96–7, 106–7
Fortnum and Mason (London shop), 73

Galloway, B. T., 26
Gardener's Chronicle, 43, 98, 168
Gedeh (volcano, Java), 16
Gentiana wardii, 30
Germany: colonies in China, 59
Giles, Professor H. A., 11
Gimson, C., 167
Gyantze (Tibet), 70

Hadfield, Eve, 116, 127
Hatton Gore (Harlington), 115
Himalayan forget-me-not, *see Chionocharis hookeri*
Himalayan mountains: K-W's observations on formation of, 41, 55
Hinks, Arthur Robert, 53, 56, 61, 76, 81, 88, 90, 93, 96
Holder, Mrs L. G., 167
Hooker, Sir Joseph: *Himalayan Journal*, 15
Hpimaw Fort (Burma), 51, 59
Huxley, T. H., 3
Hydrocharis morsus-ranae, 122

Imaw Bum (Burma), 59
Imperial Star, MV, 146
Imphal, 119–20, 158–9, 167, 170
India: K-W reaches in World War II, 152–3
Indian Botanical Survey, 169–70
Indian Tea Association, 168
Ingwersen's nurseries, West Hoathly (Sussex), 169, 175
Ingwersen, Will, 175
insects: K-W's interest in, 108–9
Iris kumaonensis (dwarf iris), 172
Irrawaddy river, 41
Ivory Poppy, 27

Japan: in World War I, 59; attack on China, 131; in World War II, 149–51, 155, 158

214

Index

Jasminum arborescens, 133
Java, 15

Kamlang Valley Expedition, 1949, 177
Kaulback, Ronald, 45, 47, 110–13
Keltie, (Sir) John Scott, 44–5, 47–9
Kew: Royal Botanic Gardens, 80, 169
Khasia-Jaintia Hills, 164, 177
Khowong Tea Estate, Assam, 163
Kingdon, Fr George (K-W's cousin), 3
Kingdon-Ward, Florinda (*née* Norman-Thompson; K-W's first wife): courtship and marriage, 65–8; background and character, 67; takes baths, 75; meets K-W in Rangoon, 98; marriage difficulties, 115–16, 118, 126, 175; children, 116; finances, 116–17; divorce, 127–31; K-W meets after divorce, 141–3; and K-W's second marriage, 169
Kingdon-Ward, Frank: birth, 1; childhood and upbringing, 2–6; relations with sister Winifred, 2; religious attitudes, 3, 5; in France, 5–6; schooling, 6, 9–11, 22; adventures with Kenneth Ward, 6–9; at Cambridge, 11; teaches in Shanghai, 11, 15, 25; on Bedford expedition in China, 16–17, 21, 24–5; as loner, 17, 113; treatment of native population and porters, 17–21, 34, 36–7, 111–12; descriptive powers, 21–3; love of rivers, 22–3; in Yunnan for Bulley, 28–30; sense of isolation, 30; and locating of plants, 30–1, 106; seed and plant collecting techniques, 30–2; medical demands on, 32–3; dogs, 40, 52; geological observations, 41–2, 55; on plant distribution, 41–2, 55; writings and literary earnings, 43–4, 98, 158; lectures, 44, 98; honoured by RGS, 44; surveying and mapping, 45–7, 61–2; carries gramophone, 50; falls in love, 52, 60; service in First World War, 53–7; saves for Tibet expedition, 54; appearance, 60, 65; opens nursery in England, 61; fevers, 62–5, 98, 123; courtship and marriage to Florinda, 65–8; expedition to Tsangpo gorges, 68, 69, 81, 89–94; plays polo, 71; slowness on expeditions, 72; provisions and diet, 73–4, 100–1, 135, 181; depressions, 75, 129, 173–4, 196–7; relations with travelling companions, 75–6, 79, 110–14; homesickness, 77; injures eye, 77; takes morphia, 78; 1926 North Burma expedition, 95; and financing of expeditions, 98; forbidding manner, 110–12; toughness, 114; marriage relations and home life, 115–16, 118, 126; children, 116–18; finances, 116–17, 138; 1935 expedition in Assam and Tibet, 119–25; plans 1937 China expedition, 127–34; divorce from Florinda, 127–31; North Burma expeditions (1937–9), 135–8; arm impaled, 135; environmental concerns, 137; in USA for New York World Fair, 138, 139–40; in London at start of World War II, 140–3; works in Censors' Office, 144–5; article on Neanderthalers, 145; posted to Far East, 146–9; World War II work in India, 151–3, 156–61; reaches Calcutta, 155; jeep accident, 160; works on Assam tea plantation, 162–5, 168; considers taking mistress, 164; social views, 165, 178; meets and corresponds with Jean (Macklin), 166, 168; searches for wrecked US aircraft, 166–7; prostate operation, 168; return to England and illness, 169; marriage to Jean, 169–70, 174–6; 1948 Manipur expedition, 169–76; 1949 Kamlang Valley Expedition, 177; spinal arthritis, 177; 1950 Lohit Valley expedition, 178–9, 181–91; nervous of heights, 182; on Assam earthquake, 183–6; plans for retirement, 193; 1952/3 North Burma expedition, 193–200; ill health, 200–1; stroke and death, 204
WORKS
Assam Adventure, 120, 122
From China to Hkamti Long, 62–3
In Farthest Burma, 51
The Land of the Blue Poppy, 29, 37, 43

Index

Kingdon-Ward, Frank—*Works (cont.)*
  Mystery Rivers of Tibet, 43
  On the Road to Tibet, 5, 18
  Plant Hunter in Manipur, 170, 172–3, 175
  A Plant Hunter in Tibet, 114
  Plant Hunter's Paradise, 106, 108
  Plant Hunting in the Wilds, 60
  Plant Hunting on the Edge of the World, 95, 97, 99–101
  The Riddle of the Tsangpo Gorges, 77, 80–1, 86, 88, 92
  The Romance of Gardening, 83
Kingdon-Ward, Jean (*née* Macklin; K-W's second wife; *later* Rasmussen): on K-W's relations with porters, 20; K-W meets and corresponds with, 166, 168; marriage, 169–70, 174–6; on Manipur expedition, 170–5; on 1950 Lohit Valley expedition, 181–3; in Assam earthquake, 184–5, 189; fevers and illnesses, 187–9, 196–7; Christian beliefs, 189; returns from Lohit expedition, 190–1; and proposed communal family home, 193; on 1953 North Burma expedition, 196–8, 200; and K-W's illness, 201; on 1956 Burma expedition, 201–3
Kingdon-Ward, Martha (K-W's daughter): born, 116; and father at home, 117; letters from K-W, 125–6, 143–4, 161; and parents' divorce, 128–9; K-W visits, 141; majority, 168; and proposed communal home, 193
Kingdon-Ward, Pleione (K-W's daughter): born, 117; and father at home, 117; letter from father, 125–6; and parents' divorce, 128–9; K-W visits, 141; as speech therapist, 176
Kuomintang, 36, 131
Kynaston, Herbert, 9

Lanning, George: *Wild Life in China*, 16
La Ti river, 186
Levinson, Constance, 40, 117
Lhasa: Norpu Lingka, 80
Lilium arboricola (Arahku lily), 195–6, 198

L. henryi, 173
L. mackliniae (earlier believed to be nomocharis), 167, 171–2, 174–5, 195
L. wardii ('Pink martagon' or Tsangpo lily), 98
Listera wardii, 30
Lohit Valley, 179, 181, 186, 189–90
Londonderry, Edith Helen, Marchioness of, 68

Macartney, George, Earl, 13
McDonnell, Michael, 10
Mackenzie, Compton: *Sinister Street*, 9
Macklin, Sir Albert Sortain Romer, 166, 169
Macklin, Jean, *see* Kingdon-Ward, Jean
Mahabaleshwar, 161
Malaya: Japanese in, 151; *see also* Singapore
Manchu dynasty: fall of, 34, 36
Mandalay, 154
Manipur, 156–7, 168–76
Maru tribe, 52
Meconopsis baileyi, see *Meconopsis betonicifolia*
M. betonicifolia (formerly *baileyi*: Tibetan blue poppy), 80–3
M. florindae, 68, 88
M. impedita var. *rubra*, 97
M. integrifolia, 42
M. wallichii, 52
M. wardii, 30
Mekong river, 38–9, 41
Mesopotamia: K-W serves in during World War I, 56–7
Microtus custos (vole), 16
M. wardii (vole), 16
Milligan (Wisley student), 100
Mishmi Hills, 177
Mishmis (porters), 111–12
Mukerjee, S. K., 170–1
Mundy (K-W's wartime assistant), 157
Myitkyina (Burma), 136, 154

Naga Hills, 119, 177
Nam Tamai, 134
Natural History Museum, London, 80, 140–1, 193

Index

Nemours (house, Delaware), 139
Ness (garden), Neston (Cheshire), 27
New York Botanic Garden, 100, 143, 168–9, 176–7
New York World Fair, 1939, 138–40
New Zealand Rhododendron Society, 178, 191, 193
nomocharis, 137, 139; *see also Lilium mackliniae*
Nomocharis nana, 122
Norway: K-W visits, 204

papaya: as 'caviar', 137
Papua New Guinea, 193
Pegu (Burma), 150
Pemako province (Tibet), 80
Penang, 133, 147; invaded, 151
Percy Sladen Memorial Fund, 98
Phari (Tibet), 69–70
pink martagon, *see Lilium wardii*
Pome (mountain range), 121
Pomed province (China), 48–9
Poona, 159
Primula agleniana var. *thearosa* (tea rose primula), 97
P. alpicola 'Joseph's Sikkimensis Primula', 87
P. florindae, 68, 87–8
P. pulchelloides, 77
P. sikkimensis 'Moonlight', 87–8
P. vernicosa, 30
Prince of Wales, HMS, 151
Prunus cerasoides rubea (carmine cherry), 105–6
Pu Yi, Emperor of China, 14, 36
Putao (Burma), 154–5
Pyrus 'Goldbeam', 109

Quercus griffithii, 133
Q. semiserrata, 133
Q. serrata, 133

Rainbow Fall (Tsangpo river), 90–1
Rangoon, 103, 133; in World War II, 148, 150, 152
Ranunculus aquaticus (water buttercup), 122
Rasmussen, Jean, *see* Kingdon-Ward, Jean

Repulse, HMS, 151
Rhododendron aganniphum, 50
R. agglutianum (*R. cawdorensis*), 85
R. bullatum, 60, 182
R. campylocarpum 'Yellow Peril', 86–7
R. campylogynum, 50
R. campylogynum 'Plum Warner', 86
R. cawdorensis, see *R. agglutianum*
R. cerasinum 'Coals of fire', 86
R. var. *chamaethauma* 'Carmelite', 86–7
R. charianthum, 50
R. dichroathum, 50
R. magnificum 'Rose Purple', 109
R. martianum, 189
R. megeratum, 60, 199
R. melianthum, 50
R. myrtilloides, 59
R. nuttallii 'Madonna', 86
R. repens 'Scarlet Pimpernel', 86–7
R. repens var. *chamaedoxa* 'Scarlet Runner', 86
R. roylei 'Orange Bill', 83–6
R. triflorum, 77
R. tsangpoense 'Plum Glaucum', 86
R. wardii, 50
Rhododendron 'Orange Bill', *see R. roylei* 'Orange Bill'
rhododendrons, 30, 50, 83–7, 122
Rima, 183
Rima Road, 152–3
Rollins, J. W., 143, 168, 176
Rosa moyesii, 191
R. sericea, 191
Rothschild, Lionel de, 98, 141
Royal Botanic Gardens, Edinburgh, *see* Edinburgh
Royal Botanic Gardens, Kew, *see* Kew
Royal Geographical Society, London, 45–7, 50, 70, 168
Royal Horticultural Society, London, 141, 167, 169, 193
Royal Society of London, 98

Sadiya (Assam), 178, 191–2
St Paul's School, London, 6, 9–10, 22
Salween river, 41
Saramati, Mount (Burma), 200

Index

Saxifraga wardii, 30
Scott Keltie, J., *see* Keltie, (Sir) John Scott
Seinghku river, 96
Shanghai, 14–15
Shanghai Mercury, 5
Shanghai Public School, 11, 15, 25
Silene rosaeflora, 30
Singapore: K-W visits, 15, 133–4; in World War II, 148–9
Sirhoi (mountain, Assam), 167, 171–2
Sorex wardii (shrew), 16
Staffordshire, S.S., 194
Stalin, Josef, 131
Stanford, J. K., 119
Stern, Colonel (Sir) Frederick C., 175
Sumpra Bum (Burma), 197, 200
Sun Yat-sen, 14, 36; death, 131
Sung Chiao-jen, 36
Sweden: K-W visits, 204

Tahawndam (Burma), 109–10
Tama Bum (mountain, Burma), 196, 199
Tashi Thondup (Tibetan servant), 19, 121
Taylor, Sir George, 27; *An Account of the Genus Meconopsis*, 81
'Tea Rose' primula, *see Primula agleniana*
T'eng Yueh (Yunnan), 40
Thailand, 148
Thalam Bum (mountain, Burma), 189
Thomas, Oldfield, 16
Thyne, Mrs 'Tiny', 105
Tibet: K-W on Bedford expedition to, 16–17; 1933 expedition to, 19, 119–25; political unrest in, 33–4; K-W's entry restricted (1914), 48–9; K-W plans 1915 expedition to, 54; K-W collects in, 80; mountains, 121; 1950 entry to, 181, 183; 1950 political situation in, 181–2; Chinese occupy, 182
Tibetan blue poppy, *see Meconopsis betonicifolia*
Tisang river, 96
Tocklai Tea Research Station, Ciunamara (Assam), 164, 176, 193
Tsai-feng, Regent of China, 14
Tsangpo gorges: 1924 expedition to, 68, 69–70, 77, 81, 89–93

Tsangpo lily, *see Lilium wardii*
Tsoga, Lake, 122–3

United States Department of Agriculture Bureau of Plant Industry, 25–6

Vernay, Arthur S., 100, 119, 121, 136
Vernay-Cutting Expedition, 1938, 136, 138
Victoria, Mount (Burma), 201

Walker, Frederick William, 9–10
Wallaces nursery (Tunbridge Wells), 169
Ward, Dorothy (K-W's deceased sister), 1
Ward, Harry Marshall (K-W's father), 1–4, 6, 28; death, 11, 43
Ward, Kenneth, 6–8, 42
Ward, Selina ('Lina') Mary (*née* Kingdom; K-W's mother), 1, 43, 60; death, 65
Ward, Winifred (K-W's sister): childhood and upbringing, 1–4; on K-W's lack of confidence, 28; K-W visits in Cambridge, 43; on K-W's girl-friend, 52; and nother's death, 65; letters from K-W, 104, 123, 126–7, 136, 146, 148–9, 156–8, 171, 175, 189, 190, 202; K-W's allowance to, 116–17; and K-W's first marriage, 118; and K-W's fevers, 123; and K-W's wartime travels, 146, 148–9; and K-W's war work, 156–8; and K-W with Jean in Manipur, 171; opens speech therapy school, 176; and proposed communal home, 193
water buttercup, *see Ranunculus aquaticus*
Waterhouse, Alfred, 9
Wavell, General Sir Archibald, 155–6
Williams, J. C., 27
wood oil tree, *see Aleurites fordii*
World War I, 53–6
World War II, 140–2, 146–56

Yangtse river, 22, 38
Yeates, J. S., 191
Younghusband, Colonel (Sir) Francis, 14
Yuan Shih-k'ai, 36, 59
Yunnan: K-W in, 28–30, 40, 48; revolution in, 34; 1921 trip to, 61–2; 1937 expedition planned, 131
Yuri, Colonel, 121–2